Achieve great Chemistry with CGP!

Let's deal with the bad news first: the new Grade 9-1 GCSE Chemistry courses are tougher than ever, so you'll need to be at the top of your game on exam day.

Here's the good news: this fantastic CGP book is absolutely jam-packed with all the exam-style practice you'll need — it even covers the new required practicals.

We've included fully-worked answers too, so it's easy to check how you're doing and find out how to pick up the best marks possible!

CGP — still the best! ☺

Our sole aim here at CGP is to produce the highest quality books — carefully written, immaculately presented and dangerously close to being funny.

Then we work our socks off to get them out to you
— at the cheapest possible prices.

Contents

✓ Use the tick boxes to check off the topics you've completed.

Topic 1 — Atomic Structure and the Periodic Table

Atoms .... 1 ☐
Elements .... 2 ☐
Compounds .... 3 ☐
Chemical Equations .... 4 ☐
Mixtures and Chromatography .... 5 ☐
More Separation Techniques .... 6 ☐
Distillation .... 7 ☐
The History of The Atom .... 8 ☐
Electronic Structure .... 10 ☐
Development of The Periodic Table ...... 11 ☐
The Modern Periodic Table .... 12 ☐
Metals and Non-Metals .... 13 ☐
Group 1 Elements .... 14 ☐
Group 7 Elements .... 15 ☐
Group O Elements .... 17 ☐

Topic 2 — Bonding, Structure and Properties of Matter

Formation of Ions .... 18 ☐
Ionic Bonding .... 19 ☐
Ionic Compounds .... 21 ☐
Covalent Bonding .... 23 ☐
Simple Molecular Substances .... 25 ☐
Polymers and Giant Covalent Substances ... 27 ☐
Allotropes of Carbon .... 28 ☐
Metallic Bonding .... 29 ☐
States of Matter .... 30 ☐
Changing State .... 31 ☐
Nanoparticles .... 32 ☐
Uses of Nanoparticles .... 34 ☐

Topic 3 — Quantitative Chemistry

Relative Formula Mass .... 35 ☐
The Mole .... 36 ☐
Conservation of Mass .... 37 ☐
The Mole and Equations .... 39 ☐
Limiting Reactants .... 41 ☐
Gases and Solutions .... 42 ☐
Concentration Calculations .... 44 ☐
Atom Economy .... 46 ☐
Percentage Yield .... 47 ☐

Topic 4 — Chemical Changes

Acids and Bases .... 48 ☐
Titrations .... 49 ☐
Strong Acids and Weak Acids .... 50 ☐
Reactions of Acids .... 51 ☐
The Reactivity Series .... 53 ☐
Separating Metals from Metal Oxides ... 55 ☐
Redox Reactions .... 56 ☐
Electrolysis .... 57 ☐

Topic 5 — Energy Changes

Exothermic and Endothermic Reactions ... 60 ☐
Bond Energies .... 62 ☐
Cells, Batteries and Fuel Cells .... 63 ☐

Topic 6 — The Rate and Extent of Chemical Change

Rates of Reaction .... 65 ☐
Measuring Rates of Reaction .... 68 ☐
Reversible Reactions .... 71 ☐
Le Chatelier's Principle .... 72 ☐

Topic 7 — Organic Chemistry

Hydrocarbons .... 74 ☐
Fractional Distillation .... 76 ☐
Uses and Cracking of Crude Oil .... 77 ☐
Alkenes .... 79 ☐
Reactions of Alkenes .... 80 ☐
Addition Polymers .... 82 ☐
Alcohols .... 83 ☐
Carboxylic Acids .... 85 ☐
Condensation Polymers .... 87 ☐
Naturally Occurring Polymers .... 88 ☐

Topic 8 — Chemical Analysis

Purity and Formulations .... 89 ☐
Paper Chromatography .... 90 ☐
Tests for Gases and Anions .... 92 ☐
Flame Emission Spectroscopy .... 94 ☐

Topic 9 — Chemistry of the Atmosphere

The Evolution of the Atmosphere .... 95 ☐
Greenhouse Gases and Climate Change ... 97 ☐
Carbon Footprints .... 99 ☐
Air Pollution .... 100 ☐

Topic 10 — Using Resources

Ceramics, Polymers and Composites 101

Properties of Materials 102

Corrosion 103

Finite and Renewable Resources 104

Reuse and Recycling 105

Life Cycle Assessments 107

Potable Water 108

Waste Water Treatment 110

The Haber Process 111

NPK Fertilisers 112

Mixed Questions

Mixed Questions 113

Answers

Answers 125

Published by CGP

Editors:
Katie Braid, Robin Flello, Emily Forsberg, Emily Howe, Sharon Keeley-Holden.

Contributors:
Ian Davis, Alison Dennis, John Duffy, Mark Ellingham, Chris Workman.

With thanks to Helen Ronan and Sophie Scott for the proofreading.

Data on the graph on page 98 showing the change in atmospheric CO_2 concentration provided by Carbon Cycle and Greenhouse Gases group, 325 Broadway R/CSD, Boulder, CO 80305 (http://esrl.noaa.gov/gmd/ccgg/)

Data on the graph on page 98 showing the change in global temperature from GISTEMP Team, 2015: GISS Surface Temperature Analysis (GISTEMP). NASA Goddard Institute for Space Studies. Dataset accessed 2016-04-11 at http://data.giss.nasa.gov/gistemp/.

Data for the level of carbon dioxide in the atmosphere in the table on page 98: Hansen, J., R. Ruedy, M. Sato, and K. Lo, 2010: Global surface temperature change, Rev. Geophys., 48, RG4004, doi:10.1029/2010RG000345.

Data for the level of methane in the atmosphere in the table on page 98: IPCC, 2013: Climate Change 2013: The Physical Science Basis. Contribution of Working Group I to the Fifth Assessment Report of the Intergovernmental Panel on Climate Change [Stocker, T. F., D. Qin, G.-K. Plattner, M. Tignor, S. K. Allen, J. Boschung, A. Nauels, Y. Xia, V. Bex and P. M. Midgley (eds.)]. Cambridge University Press, Cambridge, United Kingdom and New York, NY, USA. (From Chapter 8.3.2)

Data for the level of CFC-12 in the atmosphere in the table on page 98: IPCC, 2013: Climate Change 2013: The Physical Science Basis. Contribution of Working Group I to the Fifth Assessment Report of the Intergovernmental Panel on Climate Change [Stocker, T. F., D. Qin, G.-K. Plattner, M. Tignor, S. K. Allen, J. Boschung, A. Nauels, Y. Xia, V. Bex and P. M. Midgley (eds.)]. Cambridge University Press, Cambridge, United Kingdom and New York, NY, USA. (From Table 8.A.1)

Data for the greenhouse gas lifetimes in the atmosphere and global warming potentials in the table on page 98: IPCC, 2013: Climate Change 2013: The Physical Science Basis. Contribution of Working Group I to the Fifth Assessment Report of the Intergovernmental Panel on Climate Change [Stocker, T. F., D. Qin, G.-K. Plattner, M. Tignor, S. K. Allen, J. Boschung, A. Nauels, Y. Xia, V. Bex and P. M. Midgley (eds.)]. Cambridge University Press, Cambridge, United Kingdom and New York, NY, USA. (From Chapter 8, Table 8.A.1)

www.cgpbooks.co.uk
Clipart from Corel®
Printed by Elanders Ltd, Newcastle upon Tyne

Based on the classic CGP style created by Richard Parsons.

How to Use This Book

- Hold the book <u>upright</u>, approximately <u>50 cm</u> from your face, ensuring that the text looks like <u>this</u>, not ᴙᴉɥ⊥.
  Alternatively, place the book on a <u>horizontal</u> surface (e.g. a table or desk) and sit adjacent to the book,
  at a distance which doesn't make the text too small to read.
- In case of emergency, press the two halves of the book together <u>firmly</u> in order to close.
- Before attempting to use this book, familiarise yourself with the following <u>safety information</u>:

The questions are arranged into sub-topics, so you can get exam practice on exactly the bit of your course that you want.

The History of The Atom

There are warm-up questions for the trickier sub-topics, to ease you in and get you thinking along the right lines.

Warm-Up

For the following statements, circle whether they are **true** or **false**.

New experimental evidence can disprove models	True Or	False
Scientific models can be based on existing theories and new experimental evidence	True Or	False
Older scientific theories must be ignored when new ones are adopted	True Or	False

These grade stamps help to show how difficult the questions are. Remember, to get a top grade you need to be able to answer <u>all</u> the questions, not just the hardest ones.

3 Soluble metal salts can be made from the reactions of acids and metal oxides. (Grade 6-7) **PRACTICAL**

3.1 Zinc chloride can be made from the reaction of zinc oxide and hydrochloric acid.
Describe a laboratory method to produce pure crystals of zinc chloride using this reaction.

....

....

....

....

....

....

....

[4]

You'll have done some 'required practical activities' as part of your course, and you could be asked about any of them in your exams. Whenever one of the required practical activities crops up in this book, it's marked up like this.

In the real exams, some questions will be marked using a 'levels of response' mark scheme. In this book, we've marked these questions with an asterisk (*****). For these questions there'll be quite a few marks up for grabs, and you'll be marked on the <u>overall quality</u> of your answer. So make sure you answer the question fully, include as much detail as you can and make sure your answer is logical and coherent.

3.2 Suggest an alternative to zinc oxide that would also react with hydrochloric acid to form the desired product.

....

[1]

[Total 5 marks]

4* A student is given three different unlabelled solutions. One contains sodium hydroxide, one contains sodium carbonate and one contains a sodium salt. Use your knowledge of chemical reactions to describe experiments that the student could do to decide which solution is which. Explain how your experiments would allow the student to identify the solutions. Clearly describe the reactants and products of any reactions you include. (Grade 7-9)

....

....

....

You're told how many marks each question part is worth, and then the total for the whole question.

Exam Practice Tips give you hints to help with answering exam questions.

Exam Practice Tip

When looking at the history of the atom, it's not only important that you know the different theories of atomic structure — you also need to understand why, and describe how, scientific theories develop over time. So make sure you go through your notes on Working Scientifically before the exam. And then why not relax with some plum pudding...

☹ ☐   ☺ ☐   😊 ☐   Topic 1 — Atomic Structure and the Periodic Table

Tick the box that matches how confident you feel with the questions in each sub-topic. This should help show you where you need to focus your revision.

Atoms

Choose from the labels below to fill in the blanks in the passage.

protons

1×10^{-14}

0.1

1/10 000

charge   electrons

The radius of an atom is approximately .... nanometres.

The radius of the nucleus is around .... metres.

That's about .... of the radius of an atom.

An atom doesn't have an overall .... as it has equal

numbers of .... and ....

1 **Figure 1** shows the structure of a helium atom. (Grade 4-6)

Figure 1

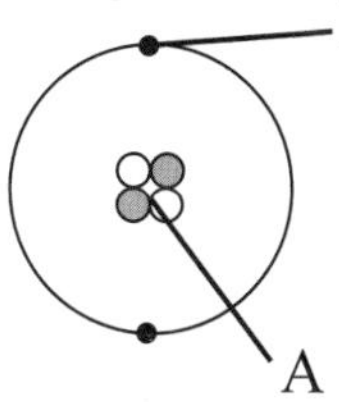

1.1 Name the region where most of the mass of the atom is concentrated.

....

[1]

1.2 What is the relative charge of particle **B**?

....

[1]

1.3 Name the particles in region **A** and give their relative charges.

....

....

[2]

[Total 4 marks]

2 A potassium atom can be represented by the nuclear symbol $^{39}_{19}$K. (Grade 4-6)

2.1 What is the mass number of $^{39}_{19}$K?

....

[1]

2.2 What is the atomic number of $^{39}_{19}$K?

....

[1]

2.3 How many protons, neutrons and electrons does an atom of $^{39}_{19}$K have?

protons: ....   neutrons: ....   electrons: ....

[3]

[Total 5 marks]

Elements

1 $^{35}_{17}Cl$ and $^{37}_{17}Cl$ are the naturally occurring isotopes of the element chlorine. (Grade 4-6)

1.1 Which of the following statements about elements is true? Tick **one** box.

Atoms of an element can contain different numbers of protons. ☐

Chlorine is one of 200 different elements. ☐

Elements contain more than one type of atom. ☐

Atoms are the smallest part of an element that can exist. ☐

[1]

1.2 Explain why $^{35}_{17}Cl$ and $^{37}_{17}Cl$ are isotopes of the element chlorine.

....

....

....

[2]

[Total 3 marks]

2 This question is about isotopes. (Grade 6-7)

2.1 A neutral atom of sulfur, ^{32}S, has 16 electrons.
Sulfur has three other naturally occurring isotopes, with mass numbers 33, 34 and 36.
Complete the table below giving the number of protons, neutrons and electrons for each isotope.

isotope	number of protons	number of neutrons	number of electrons	% abundance
^{32}S	....	....	16	94.99
^{33}S	....	....	....	0.75
^{34}S	....	....	....	4.25
^{36}S	....	....	....	0.01

[3]

2.2 Using the information in the table above, calculate the relative atomic mass of sulfur.
Give your answer to one decimal place.

Relative atomic mass: ....

[2]

2.3 Atom **X** has a mass number of 51 and an atomic number of 23.
Atom **Y** has a mass number of 51 and an atomic number of 22.
Atom **Z** has a mass number of 52 and an atomic number of 23.

Identify which of the atoms are isotopes and explain why.

....

....

[3]

[Total 8 marks]

Topic 1 — Atomic Structure and the Periodic Table

Compounds

1 Ammonia is a compound with the formula NH_3. **Grade 4-6**

1.1 Why is ammonia classified as a compound? Tick **one** box.

It contains only one type of atom. ☐

It contains two elements chemically combined. ☐

It cannot be broken down into elements using chemical methods. ☐

It contains more than one atom. ☐

[1]

1.2 How many atoms are there in a single molecule of ammonia?

....

[1]

[Total 2 marks]

2 The following list contains a variety of substances identified by their chemical formula. **Grade 6-7**

A. O_2 **B.** $NaCl$ **C.** C_2H_4 **D.** P_4 **E.** H_2O **F.** H_2

2.1 Name substance **B**.

....

[1]

2.2 Identify **one** substance from the list that is a compound. Explain your choice.

....

....

[2]

2.3 How many atoms are there in a molecule of substance **C**?

....

[1]

2.4 **C** and **F** were reacted to form C_2H_6. Using the formulas of the substances, state if a new compound has been made and explain your answer.

....

....

....

[1]

[Total 5 marks]

Exam Practice Tip

Make sure you know the difference between atoms, elements and compounds. Here's a quick summary... Everything is made of atoms (which contain protons, neutrons and electrons). Elements only contain one type of atom (all the atoms have the same number of protons). A compound consists of different atoms joined together by chemical bonds. Got it?

        Topic 1 — Atomic Structure and the Periodic Table

Chemical Equations

Warm-Up

The chemical word equation for a reaction is shown below:

magnesium + hydrochloric acid → magnesium chloride + hydrogen

For each of the following statements circle whether the statement is **true** or **false**.

1)	Hydrogen is a product in the reaction	True Or False
2)	The equation shows the reaction between chlorine and hydrogen	True Or False
3)	Hydrochloric acid is a reactant	True Or False
4)	The equation shows the reaction between magnesium and hydrochloric acid	True Or False

1   Sodium (Na) reacts vigorously with chlorine gas (Cl_2) to form sodium chloride (NaCl) only.   *Grade 4-6*

1.1   Write a word equation for this reaction.

....

[1]

1.2   Which of the following equations correctly represents this reaction?

$Na + Cl \rightarrow NaCl$ ☐

$Na_2 + 2Cl \rightarrow 2NaCl$ ☐

$Na_2 + Cl_2 \rightarrow 2NaCl$ ☐

$2Na + Cl_2 \rightarrow 2NaCl$ ☐

[1]

[Total 2 marks]

2   Nitric acid can be made using ammonia.   *Grade 7-9*

2.1   The first stage in the manufacture of nitric acid is to oxidise ammonia, NH_3, to nitrogen(II) oxide, NO. Balance the equation for the reaction.

.... NH_3 + ... O_2 → .... NO + ... H_2O

[1]

2.2   The reaction below shows the final stage in the manufacture of nitric acid. The equation is not balanced correctly. Explain how you can tell.

$$2NO_2 + O_2 + H_2O \rightarrow 2HNO_3$$

....

....

[1]

[Total 2 marks]

Topic 1 — Atomic Structure and the Periodic Table

Mixtures and Chromatography

1 Air contains many gases. It mainly consists of nitrogen, N_2, and oxygen, O_2, with a small amount of carbon dioxide, CO_2. There are also trace amounts of noble gases such as argon, Ar.

Grade 4-6

1.1 Is air an element, compound or mixture? Explain your answer.

....

....

....

[3]

1.2 Argon can be separated out from air. Will the chemical properties of argon as a separate gas be different from the properties of argon in air? Explain your answer.

....

....

[2]

[Total 5 marks]

2* A simple chromatography experiment was set up to separate dyes in inks. Samples of three different water soluble inks, A, B and C, were placed on a piece of chromatography paper. After 30 minutes, the chromatogram in **Figure 1** was obtained.

Grade 6-7

Figure 1

PRACTICAL

Solvent front

Solvent origin

A B C

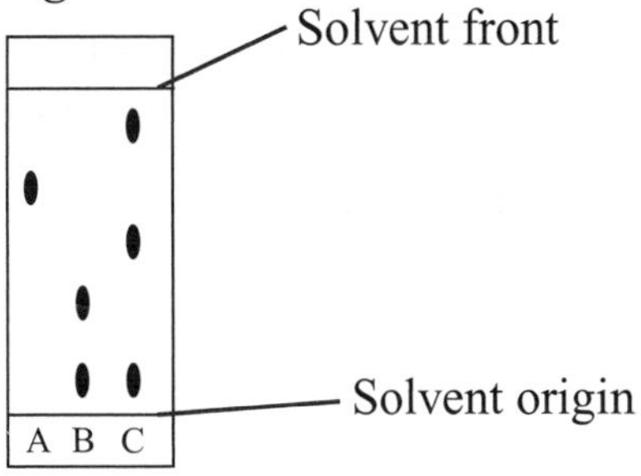

Outline the procedure for setting up and running this experiment and explain the results.

....

....

....

....

....

....

....

....

[Total 6 marks]

Topic 1 — Atomic Structure and the Periodic Table

More Separation Techniques

1 A student was given a mixture of insoluble chalk powder and solid potassium chloride. Potassium chloride is water soluble.
Grade 6-7

1.1 Outline a method for separating the chalk from potassium chloride.

....

....

....

[3]

1.2 Outline the process the student could use to obtain pure crystals of potassium chloride from a potassium chloride solution.

....

....

....

[1]

[Total 4 marks]

2 Powdered iron and powdered sulfur were mixed together.
Information on the solubility of iron and sulfur is given in **Table 1**.
Grade 7-9

Table 1

	Iron	Sulfur
Solubility in water	Insoluble	Insoluble
Solubility in methylbenzene	Insoluble	Soluble

2.1 Using the information in **Table 1**, explain how the iron and sulfur could be separated out again to form pure iron and crystals of sulfur.

....

....

....

....

....

[3]

2.2 Some of the mixture was heated in a fume cupboard. Iron(II) sulfide formed.
A student stated the iron and sulfur could be separated back out using physical processes.
Is the student correct? Explain your answer.

....

....

....

[3]

[Total 6 marks]

Topic 1 — Atomic Structure and the Periodic Table

Distillation

1 Sucrose is a sugar with the formula $C_{12}H_{22}O_{11}$. It is soluble in water.

Grade 4-6

Which of the following techniques would be best for obtaining pure water from a sugar solution? Tick **one** box.

☐ Evaporation ☐ Condensation ☐ Simple distillation ☐ Fractional distillation

[Total 1 mark]

2 A solution containing aspirin, which has a boiling point of 140 °C, was prepared. The solution contained an impurity with a boiling point of 187 °C. The distillation apparatus shown in **Figure 1** was set up to separate the aspirin from the impurity.

Grade 6-7

Figure 1

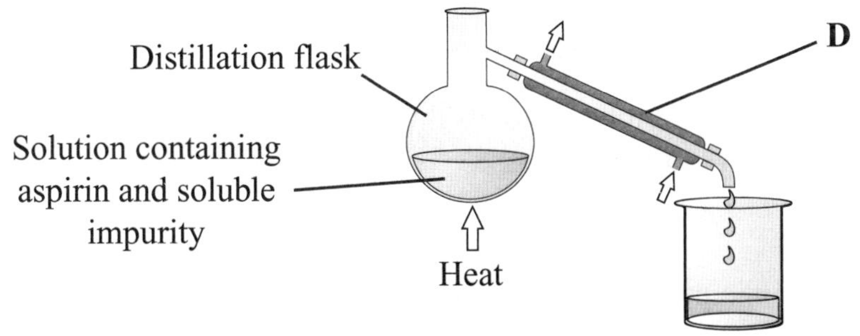

Distillation flask

Solution containing aspirin and soluble impurity

Heat

D

2.1 The apparatus is not correctly set up.
Explain how you would modify the apparatus to make it function correctly.

....

....

[1]

2.2 Explain how aspirin is separated from the impurity when the distillation apparatus is set up correctly. Include the function of D in your answer.

....

....

....

....

[3]

2.3 If the distillation flask was placed in a boiling water bath and not directly heated, the aspirin would not distill. Explain why.

....

....

[2]

[Total 6 marks]

The History of The Atom

For the following statements, circle whether they are **true** or **false**.

New experimental evidence can disprove models	True Or False	
Scientific models can be based on existing theories and new experimental evidence	True Or False	
Older scientific theories must be ignored when new ones are adopted	True Or False	

1 Models of the atom have changed over time. (Grade 4-6)

1.1 Which of the following statements is the best description of what scientists thought an atom was like before the electron was discovered?
Tick **one** box.

☐ Tiny solid spheres that can't be divided.   ☐ Formless 'clouds' of matter.   ☐ Flat geometric shapes.   ☐ Discrete packets of energy.

[1]

1.2 Draw one line from each atomic model to the correct description of that model.

Atomic Model   **Description**

Plum pudding model

A positively charged 'ball' with negatively charged electrons in it.

A small positively charged nucleus surrounded by a 'cloud' of negative electrons.

Bohr's model

Electrons in fixed orbits surrounding a small positively charged nucleus.

Rutherford's nuclear model   Solid spheres with a different sphere for each element.

[3]

1.3 In 1932, James Chadwick discovered a neutral sub-atomic particle found in the nucleus. Give the name of this particle.

....

[1]

[Total 5 marks]

2 In 1911, Ernest Rutherford fired alpha particles at gold foil. Most of the alpha particles passed straight through the gold foil without being deflected. Only a few of the particles bounced back at 180°.

Grade 6-7

2.1 Explain why most of the alpha particles passed through the foil.

....

....

[1]

2.2 Name the scientist that adapted Rutherford's nuclear model by suggesting that electrons orbit the nucleus at specific distances.

....

[1]

[Total 2 marks]

3 In 1897, J. J. Thomson discovered the electron. This discovery led Thomson to propose the 'plum pudding' model of the atom.

Grade 7-9

3.1 Thomson realised that there must be positive charge in an atom as well as the negative charge of the electrons. Suggest how Thomson could have worked this out using his knowledge of the atom and its charge.

....

....

[2]

3.2* Describe similarities and differences between the plum pudding model and the modern nuclear model of the atom.

....

....

....

....

....

....

....

....

....

....

[6]

[Total 8 marks]

Exam Practice Tip

When looking at the history of the atom, it's not only important that you know the different theories of atomic structure — you also need to understand why, and describe how, scientific theories develop over time. So make sure you go through your notes on Working Scientifically before the exam. And then why not relax with some plum pudding...

Topic 1 — Atomic Structure and the Periodic Table

Electronic Structure

1   Calcium is a reactive metal with an atomic number of 20.   *(Grade 4-6)*

1.1   What is the electron configuration of a neutral calcium atom? Tick **one** box.

☐   2, 18   ☐   2, 16, 2   ☐   2, 8, 8, 2   ☐   2, 2, 8, 8

[1]

1.2   Both of calcium's electrons are in its innermost shell. Explain why this is.

....

....

[2]

[Total 3 marks]

2   Electronic structures can be represented in different ways.   *(Grade 6-7)*

Figure 1

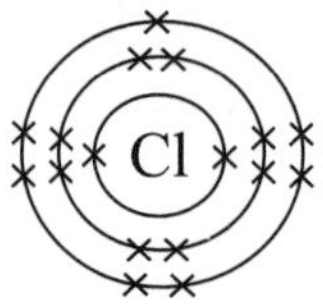

2.1   **Figure 1** shows the electronic structure of an atom of chlorine.
Give the electronic structure of chlorine in number form.

Chlorine: ....

[1]

2.2   Draw the electronic structure of sulfur.

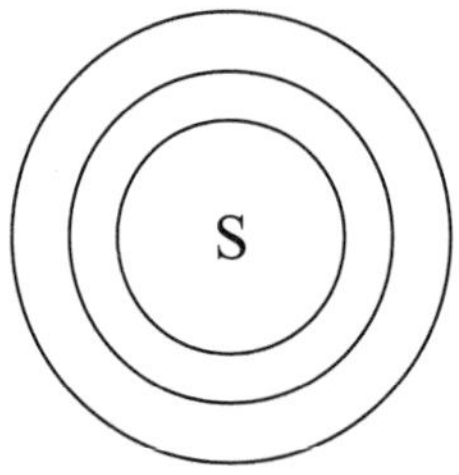

[2]

2.3   **Figure 2** shows the electronic structure of an atom of an element, **X**. Identify element **X**.

Figure 2

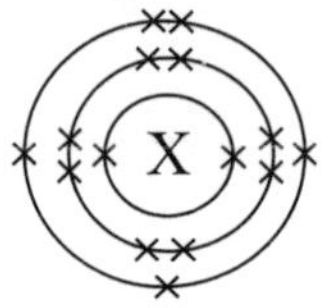

Element **X**: ....

[1]

[Total 6 marks]

Exam Practice Tip

Electronic structures are the key to understanding lots of chemistry, so it's really important that you know how to work them out. Definitely make sure you know, or can work out, the electronic structures for the first 20 elements in the periodic table before going into your exam. It'll help you in all manner of questions, believe me.

Topic 1 — Atomic Structure and the Periodic Table

Development of The Periodic Table

1 Mendeleev, guided by atomic masses, grouped elements with similar properties to form an early version of the periodic table. He used his table to predict properties of undiscovered elements such as eka-silicon, now called germanium.
Table 1 shows some predictions.

Table 1

	Silicon (Si)	Eka-silicon (Ek)	Tin (Sn)
Atomic Mass	28	72	119
Density in g/cm³	2.3	**D**	7.3
Appearance	grey/silver non-metal	grey metal	grey metal
Formula of oxide	SiO_2	**E**	SnO_2
Formula of chloride	$SiCl_4$	**F**	$SnCl_4$
Reaction with acid	None	**G**	Slow

1.1 Mendeleev left gaps in his periodic table. Suggest why.

....

....

[1]

1.2 Predict properties **D** to **G**.

D .... **E** ....

F .... **G** ....

[4]

[Total 5 marks]

2 Before the development of the modern periodic table, scientists classified elements according to their atomic masses. *(Grade 7-9)*

2.1 Why did early scientists use atomic mass and not atomic number?

....

....

[1]

2.2 Use the periodic table to identify two elements that would be in the wrong positions if the elements were ordered by atomic mass rather than atomic number.

....

[1]

2.3 How did the discovery of isotopes explain why atomic weights do not always give the correct order of elements?

....

....

[2]

[Total 4 marks]

    Topic 1 — Atomic Structure and the Periodic Table

The Modern Periodic Table

1 Chemical elements are arranged in the periodic table. (Grade 4-6)

1.1 How are the elements ordered in the periodic table?

....

[1]

1.2 Why is the table called the periodic table?

....

[1]

1.3 Why do elements in groups have similar chemical properties? Tick **one** box.

They have the same number of shells of electrons. ☐

They have the same number of outer shell electrons. ☐

They all have at least one full inner shell of electrons. ☐

The atoms of the elements are similar in size. ☐

[1]

[Total 3 marks]

2 **Figure 1** below shows the electronic configuration for a neutral atom of element **X**. (Grade 6-7)

Figure 1

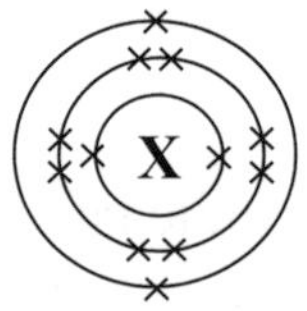

2.1 What group in the periodic table is element **X** in? Justify your answer.

....

....

[2]

2.2 Which period is element **X** in? Explain your answer.

....

....

[2]

2.3 Identify element **X**.

....

[1]

2.4 Name one element that will have similar chemical properties to element **X**.

....

[1]

[Total 6 marks]

Topic 1 — Atomic Structure and the Periodic Table

Metals and Non-Metals

1 Metals make up about 80% of all the elements in the periodic table. Aluminium and magnesium are both metals. `Grade 4-6`

1.1 Describe where metals and non-metals can be found in the periodic table.

Metals: ....

Non-metals: ....

[1]

1.2 Which of the statements below is true? Tick **one** box.

Elements that react to form negative ions are metals. ☐

Elements that react to form positive ions are metals. ☐

Elements that react to form positive ions are non-metals. ☐

Elements that do not form ions are metals. ☐

[1]

1.3 Suggest one physical property that magnesium and aluminium have in common.

....

[1]

1.4 Explain why magnesium and aluminium react in a similar way.

....

....

[2]

[Total 5 marks]

2 Iron is a transition metal. It can react with sulfur, a non-metal, to form iron sulfide. `Grade 6-7`

2.1 Using your knowledge of metals and non-metals, compare
the physical properties of iron and sulfur.

....

....

....

[4]

2.2 Many transition elements can form coloured compounds.
Give **two** other properties that are specific to transition metals.

....

[2]

2.3 Name **two** transition metals, other than iron.

....

[2]

[Total 8 marks]

Topic 1 — Atomic Structure and the Periodic Table

Group 1 Elements

1   Data for two alkali metals, **X** and **Y**, is shown in the table below.

	Melting Point / °C	Density / kg per m³
X	39	1532
Y	98	968

1.1 Given that melting point decreases as you go down Group 1, which metal has the lowest atomic number, **X** or **Y**? Justify your answer.

....

....

[2]

1.2 Construct a balanced equation for the reaction between **X** and water to form the hydroxide, **XOH**, and a gas.

....

[2]

1.3 **XOH** is soluble in water. State the pH of the solution formed: ....

[1]

[Total 6 marks]

2   The element francium is at the bottom of Group 1 in the periodic table.

2.1 Complete **Table 1** by predicting the radius and boiling point of francium. Use the data in the table to help you.

Table 1

	Boiling Point / °C	Radius of atom / pm
Rb	687.8	248
Cs	670.8	265
Fr	....	....

[2]

2.2 Compare the reactivity of francium to caesium. Justify your answer using your knowledge of francium's electron arrangement.

....

....

....

[3]

2.3 Alkali metals react with phosphorus, P_4, to form ionic phosphides. Lithium reacts with phosphorus to form lithium phosphide Li_3P. Predict the formula of francium phosphide and write a balanced equation for the reaction of phosphorus and francium.

Formula: ....

Equation: ....

[3]

[Total 8 marks]

Topic 1 — Atomic Structure and the Periodic Table

Group 7 Elements

Write the following Group 7 elements in order of increasing boiling point.

Iodine

Bromine

Chlorine

Fluorine

....

....

....

....

1 The elements in Group 7 of the periodic table are known as the halogens. *(Grade 4-6)*

1.1 Which of the following statements about the halogens is true? Tick **one** box.

They are non-metals that exist as single atoms. ☐

They are metals that exist as single atoms. ☐

They are non-metals that exist as molecules of two atoms. ☐

They are metals that exist as molecules of two atoms. ☐

[1]

1.2 Compare the chemical reactivity of chlorine and bromine. Explain your answer.

....

....

....

[3]

1.3 Halogens can react with other elements to form molecular compounds. Of the following elements, suggest which one might form a molecular compound with chlorine. Tick **one** box.

Ca ☐   Cr ☐   Na ☐   P ☐

[1]

[Total 5 marks]

2 Halogens react with many metals to form metal halides. *(Grade 6-7)*
For instance, iron reacts with bromine to form iron bromide.

2.1 Complete and balance the following symbol equation for this reaction.

................Fe + ... $\rightarrow$ ....$FeBr_3$

[2]

2.2 Iron bromide is ionic. What is the charge on the bromide ion?

....

[1]

[Total 3 marks]

Topic 1 — Atomic Structure and the Periodic Table

3 Chlorine water was added to potassium bromide solution in a test tube and the contents shaken. *(Grade 6-7)*

3.1 Complete the word equation below.

Chlorine + Potassium bromide → .... + ...

[1]

3.2 What would you observe when the two reactants are mixed?

....

[1]

3.3 Give the name for this type of reaction.

....

[1]

3.4 Will chlorine water react with potassium fluoride solution? Explain your answer.

....

....

[2]

[Total 5 marks]

4 This question is about the halogens. *(Grade 7-9)*

4.1 Describe the electronic structure of the halogens and how it changes down Group 7.

....

....

....

[2]

4.2 Hydrogen gas reacts explosively with fluorine, even at low temperatures, to form hydrogen fluoride, HF. Predict and explain how astatine might react with hydrogen. Include a balanced equation in your answer.

....

....

....

....

....

....

....

[5]

[Total 7 marks]

Exam Practice Tip

One of the most important things to learn about Group 7 elements is the trends you find as you go down or up the group. And you need to be able to explain these trends using the electronic structure of the halogens. Smashing.

Topic 1 — Atomic Structure and the Periodic Table

Group 0 Elements

1 The noble gases can be found in Group 0 of the periodic table.

1.1 Using the information in Table 1, complete the table by predicting values for the boiling point of radon, the density of xenon and the atomic radius of argon.

Table 1

Element	Boiling Point / °C	Density / g/cm³	Atomic radius / pm
Ar	−186	0.0018	....
Kr	−152	0.0037	109
Xe	−108	....	130
Rn	....	0.0097	136

[3]

1.2 Explain the chemical reactivity of krypton in terms of electron configuration.

....

....

[2]

1.3 What is the difference between the electron configuration of helium and the rest of the noble gases?

....

....

[1]

[Total 6 marks]

2 Until 1962, no noble gas compounds existed. Since then, several noble gas compounds have been made. For instance, xenon difluoride exists as white crystals which are stable at room temperature in a dry atmosphere.

2.1 Why were chemists surprised that a noble gas compound could be made?

....

[1]

2.2 Why is it unlikely that iodine would form a compound with xenon?

....

[1]

2.3 A liquid containing a mixture of neon and xenon was cooled down. One gas solidified at −249 °C and the other at −112 °C. Identify which noble gas solidified at −249 °C and which at −112 °C. Justify your answer.

....

....

....

[3]

[Total 5 marks]

Topic 1 — Atomic Structure and the Periodic Table

Formation of Ions

1   This question is about atoms forming ions by losing or gaining electrons.   Grade **4-6**

1.1   Which of the following statements about the atoms of metallic elements is correct?
Tick **one** box.

☐ Metal atoms usually lose electrons to become negative ions.

☐ Metal atoms usually gain electrons to become negative ions.

☐ Metal atoms usually gain electrons to become positive ions.

☐ Metal atoms usually lose electrons to become positive ions.

[1]

1.2   The diagram below shows the ions of four elements and four descriptions about where the elements can be found in the periodic table.
Draw lines to match each of the four ions to their description.

A^+		A non-metal from Group 6
D^-		A metal from Group 2
X^{2+}		A metal from Group 1
Z^{2-}		A non-metal from Group 7

[2]

[Total 3 marks]

2   This question is about sulfur atoms and sulfur ions.   Grade **6-7**

2.1   Sulfur is in Group 6 of the periodic table. What is the charge on a sulfur ion?

....

[1]

2.2   Sulfur atoms have the electronic structure 2,8,6. Write out the electronic structure for a sulfur ion.
Explain your answer.

....

....

....

[3]

2.3   Name the element that has atoms with the same electronic structure as a sulfur ion.

....

[1]

[Total 5 marks]

Ionic Bonding

Warm-Up

Choose from the formulas on the left to complete the table showing the dot and cross diagrams and formulas of various ionic compounds. You won't need to use all the formulas.

NaCl   MgCl$_2$ MgCl

Na$_2$O   NaO   NaCl$_2$

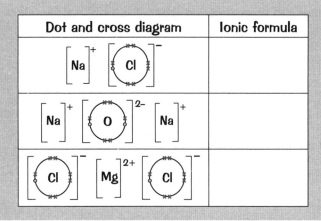

Dot and cross diagram	Ionic formula

1 Ionic bonding is one of the three types of chemical bonds found in compounds. **Grade 4-6**

1.1 In which of the following compounds are the particles held together by ionic bonds?
Put a tick in the boxes next to the **two** compounds that you think are ionic.

☐ calcium chloride   ☐ carbon dioxide   ☐ phosphorus trichloride

☐ potassium oxide   ☐ nitrogen monoxide

[2]

1.2 The dot and cross diagram below shows the formation of lithium fluoride from its elements.
The diagram is incomplete. Complete the diagram by adding an arrow to show the transfer of electron(s), the charges of the ions and completing the outer shell electronic structure of the fluoride ion.

 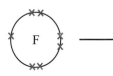

[3]

1.3 Name the force that holds the ions together in an ionic bond.

....

[1]

1.4 State how you can tell from a dot and cross diagram that the particles in a compound are held together by ionic bonds.

....

[1]

[Total 7 marks]

2 **Figure 1** shows the outer electronic structure of magnesium and oxygen.

Figure 1

2.1 Draw a similar diagram to show the electronic structures and charges of the ions that form when magnesium reacts with oxygen. You only need to show the outer shells of electrons.

[4]

2.2 Explain how an ionic bond forms when magnesium atoms react with oxygen atoms.

....

....

....

....

[4]

[Total 8 marks]

3 **Figure 2** is a representation of the structure of a compound formed from two unknown elements given the symbols X and Z.

Figure 2

$$\left[\begin{array}{c} X \end{array} \right]^{-} \left[\begin{array}{c} Z \end{array} \right]^{2+} \left[\begin{array}{c} X \end{array} \right]^{-}$$

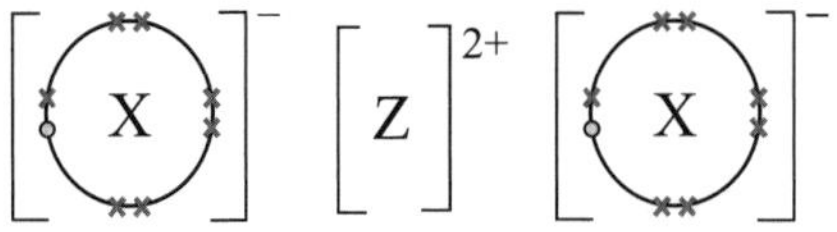

3.1 Suggest which group in the periodic table each element is from and give a reason for your choice.

Element X: Group: ....

Reason: ....

Element Z: Group: ....

Reason: ....

[4]

3.2* Discuss the uses and limitations of dot and cross diagrams.

....

....

....

....

....

....

[6]

[Total 10 marks]

Topic 2 — Bonding, Structure and Properties of Matter

Ionic Compounds

Warm-Up

Circle the correct words or phrases below so that the statement is correct.

In an ionic compound, the particles are held together by <u>weak</u>/<u>strong</u> forces of attraction. These forces act <u>in all directions</u>/<u>in one particular direction</u> which results in the particles bonding together to form <u>giant lattices</u>/<u>small molecules</u>.

1   This question is about the structure and properties of ionic compounds.   *Grade 4-6*

1.1   Which of the following properties is not typical for an ionic compound?
Tick one box.

☐ high melting and boiling points   ☐ soluble in water

☐ conduct electricity in the liquid state   ☐ conduct electricity in the solid state

[1]

1.2   Name the type of structure that ionic compounds have.

....

[1]

[Total 2 marks]

2   Sodium chloride is an ionic compound.   *Grade 6-7*

2.1   Describe the structure of a crystal of sodium chloride. You should state:
- What particles are present in the crystal.
- How these particles are arranged.
- What holds the particles together.

....

....

....

....

[4]

2.2   Explain why sodium chloride has a high melting point.

....

....

[2]

[Total 6 marks]

3 Potassium bromide has a lattice structure that is similar to sodium chloride.

3.1 Complete the diagram below to show the position and charge of the ions in potassium bromide. Write a symbol in each blank circle to show whether it is a potassium ion or a bromide ion.

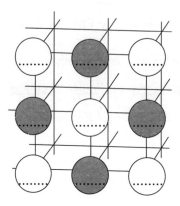

[3]

3.2 Give one advantage and one disadvantage of using the type of diagram above to represent the structure of an ionic compound.

Advantage: ....

Disadvantage: ....

[2]

3.3 What is the empirical formula of potassium bromide?

....

[1]

3.4 **Table 1** shows some data about potassium bromide. For each of the three properties shown in **Table 1**, explain how the structure of potassium bromide causes the particular property.

Table 1

Boiling point / °C	Electrical conductivity of solid	Electrical conductivity of solution
1435	Low	High

Boiling point ....

....

Electrical conductivity of solid ....

....

Electrical conductivity of solution ....

....

[6]

[Total 12 marks]

Exam Practice Tip

Don't panic if you're asked about an ionic compound that you've not met before. Think about what you _do_ know about ionic compounds, and read the question carefully to make sure you've picked up any extra information you've been given.

Topic 2 — Bonding, Structure and Properties of Matter

Covalent Bonding

1 Some elements and compounds consist of molecules that are held together by covalent bonds.

1.1 Describe how two atoms come together to form a single covalent bond.

....

[1]

1.2 What type of elements are able to form covalent bonds?

....

[1]

1.3 Borane is a compound that contains covalent bonds.
Use the diagram below to find the molecular formula of borane.

Diagram:
$$H - B - H$$
with H above B

Formula: ....

[1]

[Total 3 marks]

2 The diagrams below show dot and cross diagrams of some simple covalent molecules. Draw out the displayed formulas of these molecules using straight lines to represent covalent bonds. One molecule (H_2) has been done as an example.

Dot and cross diagram	Displayed formula
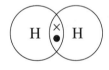	H — H
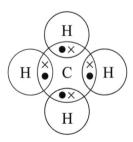	
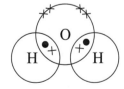	
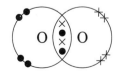	

[Total 3 marks]

Topic 2 — Bonding, Structure and Properties of Matter

3 Methane, CH_4, is a covalent molecule. The structure of methane can be shown in a number of different ways. Two such diagrams are shown in **Figures 1** and **2**.

Figure 1

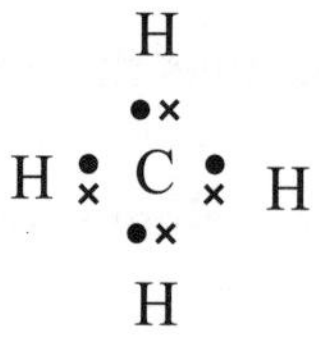

$$H \overset{\bullet\times}{\underset{\bullet\times}{\overset{\displaystyle H}{\times}}} C \overset{\bullet}{\underset{\times}{\bullet}} H$$

$$H$$

Figure 2

$$\begin{array}{c} H \\ | \\ H-C-H \\ | \\ H \end{array}$$

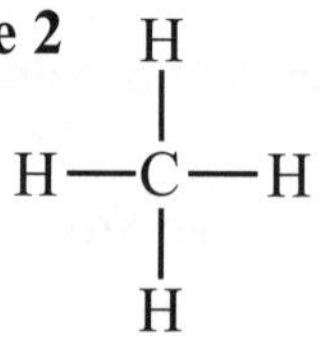

3.1 Suggest **two** different ways of telling, either from **Figures 1** and **2**, or from the molecular formula, that methane is a covalent molecule.

....

....

[2]

3.2 State **two** ways in which **Figures 1** and **2** do **not** represent a methane molecule accurately.

....

....

[2]

3.3 Briefly describe how the hydrogen atoms in a methane molecule are bonded to the carbon atom.

....

....

....

[3]

[Total 7 marks]

4 The structure of a covalent compound can be represented in many different ways.

4.1 Give one advantage and one disadvantage of using each of the representations of covalent molecules described below.

Displayed formula: ....

....

Dot and cross: ....

....

3D Model: ....

....

[6]

4.2 Suggest, with reasoning, which of the representations in 4.1 would be most suitable for drawing the structure of a large polymer in order to show what each atom is connected to.

....

....

....

[2]

[Total 8 marks]

Topic 2 — Bonding, Structure and Properties of Matter

Simple Molecular Substances

1   This question is about the forces in simple molecular substances.   (Grade 4-6)

1.1   Compare the strength of the bonds that hold the atoms in a molecule together with the forces that exist between different molecules.

....

....

[2]

1.2   When a simple molecular substance melts, is it the bonds between atoms or the forces between molecules that are broken?

....

[1]

[Total 3 marks]

2   HCl and N_2 are both simple molecular substances.   (Grade 6-7)

2.1   Draw a dot and cross diagram to show the bonding in a molecule of HCl. Show all of the outer shell electrons and use different symbols for electrons from different atoms. There is no need to show inner shell electrons.

[2]

2.2   Draw a dot and cross diagram to show the bonding in a molecule of N_2. Show all of the outer shell electrons and use different symbols for electrons from different atoms. There is no need to show inner shell electrons.

[2]

2.3   State **one** difference between the bonding in HCl compared to N_2.

....

....

[1]

[Total 5 marks]

3 Iodine, I_2, is a simple molecular substance. **Grade 6-7**

3.1 At room temperature, iodine is a solid. Explain, with reference to the forces between molecules, why this is unusual for a simple molecular substance.

....

....

[2]

3.2 Predict, with reasoning, whether iodine can conduct electricity in any state.

....

....

[2]

[Total 4 marks]

4 Both methane (CH_4) and butane (C_4H_{10}) are simple covalent compounds that are gases at room temperature. Methane has a lower boiling point than butane. **Grade 7-9**

4.1 Explain, in terms of particles, what happens when methane boils and why the boiling point of methane is lower than that of butane.

....

....

....

....

....

....

[5]

4.2 Explain why a carbon atom can form up to four covalent bonds, whilst a hydrogen atom only ever forms one covalent bond.

....

....

....

[2]

4.3 Suggest how many covalent bonds an atom of silicon would form. Explain your answer.

....

....

[2]

[Total 9 marks]

Exam Practice Tip

Each atom in a molecule should have made enough covalent bonds to get a full outer shell. So to check that you've drawn your dot and cross diagrams correctly, count up how many electrons there are in the outer shell. Unless it's hydrogen, there should be eight electrons in the outer shell. Hydrogen should end up with two electrons in its outer shell.

Topic 2 — Bonding, Structure and Properties of Matter

Polymers and Giant Covalent Substances

Warm-Up

Circle the diagram below that represents a compound with a giant covalent structure.

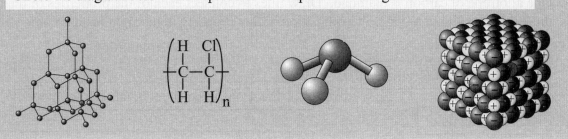

1 Graphite and diamond are compounds with very high melting points. **Grade 4-6**

1.1 Which of the following compounds is not an example of a giant covalent structure?
Tick **one** box.

☐ Ammonia ☐ Diamond ☐ Graphite ☐ Silicon dioxide

[1]

1.2 To melt a giant covalent compound, the covalent bonds between atoms must be broken.
Explain why this causes giant covalent compounds to have very high melting points.

....

....

[2]

[Total 3 marks]

2 **Figure 1** below represents a large molecule known as a polymer. **Grade 6-7**

Figure 1

$$\left(\begin{array}{cc} \text{H} & \text{H} \\ | & | \\ \text{—C—C—} \\ | & | \\ \text{H} & \text{H} \end{array}\right)_n$$

2.1 What is the molecular formula of this polymer?

....

[1]

2.2 Is this molecule likely to be a solid, liquid or gas at room temperature? Explain your answer.

....

....

....

[3]

2.3 State what type of bonds hold the atoms in the polymer together.

....

[1]

[Total 5 marks]

28

Allotropes of Carbon

1 Carbon forms a number of different allotropes. **Grade 4-6**

1.1 One allotrope of carbon is diamond. Draw lines to connect each of the properties of diamond in the left hand column to the best explanation of that property in the right hand column.

Property   **Explanation**

| Does not conduct electricity |

                                   | Electrons in covalent bonds cannot move. |

| High melting point |

                            | Each carbon atom makes four strong covalent bonds. |

| Hard (doesn't scratch easily) |

[2]

1.2 Name each of the allotropes of carbon, A-C, shown in **Figure 1**.

Figure 1

A   B 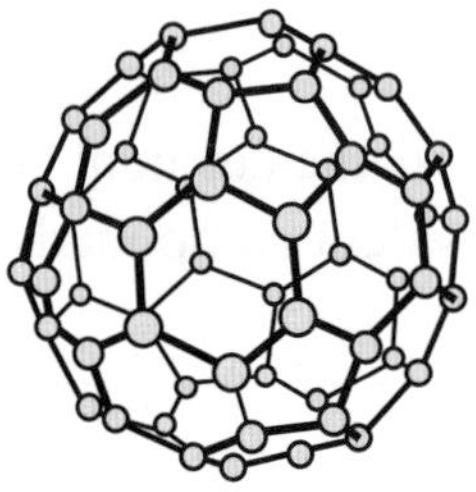   C

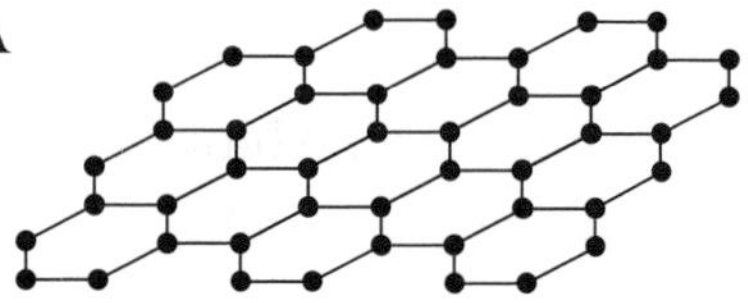

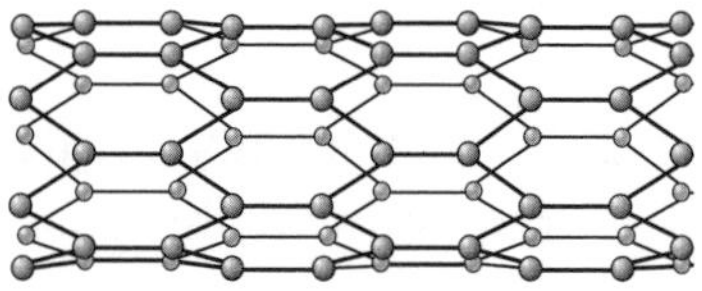

A ....   B ....   C ....

[3]

1.3 Suggest a use for allotrope C.

....

[1]

[Total 6 marks]

2 This question is about the carbon allotrope graphite. **Grade 6-7**

2.1 Describe the structure and bonding in graphite.

....

....

....

....

[4]

2.2 Graphite can be used to make components in electrical circuits.
Explain how the bonding and structure of graphite makes it suitable for this use.

....

....

[2]

[Total 6 marks]

Topic 2 — Bonding, Structure and Properties of Matter

Metallic Bonding

1 This question is about how the structure and bonding of metals affects their properties.

1.1 Draw a labelled diagram to show how the metal particles and the electrons that take part in bonding are arranged in a metal.

[3]

1.2 Explain how the atoms are held together in this arrangement.

....

....
[2]

1.3 Explain whether this arrangement causes metals to generally have high or low boiling points.

....

....
[2]

1.4 Explain whether this arrangement causes metals to be good or poor conductors of electricity.

....

....
[2]
[Total 9 marks]

2 Metals are able to be bent and shaped.

2.1 Explain how the structure and bonding in metals means they are able to be bent and shaped.

....

....
[2]

2.2 Alloys are mixtures of metals and a different element.
Explain why it is usually easier to change the shape of a pure metal than an alloy.

....

....

....
[3]
[Total 5 marks]

States of Matter

1 All substances can exist in three states of matter. **Grade 4-6**

1.1 Name the three states of matter.

....
[1]

1.2 Write the state of each substance next to each of the formulas below.

$NaCl_{(s)}$ ....

$O_{2(g)}$ ....

$Hg_{(l)}$ ....
[3]

[Total 4 marks]

2 Particle theory is a model used to explain how particles behave in the three states of matter. **Grade 6-7**

2.1 Describe how the particles in a substance are represented in particle theory.

....
[1]

2.2 Name the state of matter shown in **Figure 1**.

Figure 1

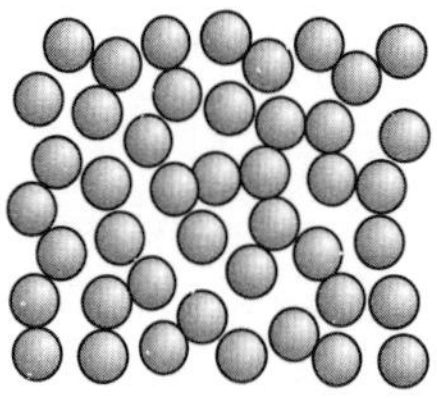

State: ....
[1]

2.3 Name two physical processes that particle theory can be used to explain.

....
[2]

2.4 Explain two ways in which particle theory doesn't accurately represent the particles in a substance.

....

....
[2]

[Total 6 marks]

Topic 2 — Bonding, Structure and Properties of Matter

Changing State

1 This question is about the processes by which a material changes state. Grade 4-6

1.1 What is the name of the process when a solid becomes a liquid?

....

[1]

1.2 What is the name of the temperature at which a liquid becomes a gas?

....

[1]

1.3 If a liquid turns into a gas at a very high temperature, what does this imply about the strength of the bonds between the particles in the substance?

....

[1]

[Total 3 marks]

2 Use the data in **Table 1** to help you answer the questions that follow: Grade 7-9

Table 1

Substance	Sodium Chloride	Water	Copper
Melting Point (°C)	801	0	1083
Boiling Point (°C)	1413	100	2567

2.1 Which substance in **Table 1** would be a liquid at 900 °C?

....

[1]

2.2 Which two substances would be gases at 1500 °C?

....

[2]

2.3 Which process requires the most energy: melting copper or boiling sodium chloride?

....

[1]

2.4 Does the data in **Table 1** suggest that the covalent bonds in a water molecule are weaker than the metallic bonds in copper? Explain your answer.

....

....

....

....

[4]

[Total 8 marks]

Topic 2 — Bonding, Structure and Properties of Matter

Nanoparticles

For each of the following statements, circle whether they are **true** or **false**.

1) Coarse particles have larger diameters than fine particles.   True or False

2) Nanoparticles have larger diameters than fine particles.   True or False

3) Atoms have smaller diameters than nanoparticles.   True or False

4) A particle with a diameter of 1000 nm would be classed as a fine particle.   True or False

5) Fine particles can also be called dust.   True or False

---

1 Particles of matter can be sorted into categories based on how large they are. Draw lines to link each of the following statements on the left-hand side with the correct number on the right-hand side.

The maximum size, in nanometres (nm), of a nanoparticle.   5000

An approximate number of atoms present in one nanoparticle.   500

The possible size, in nanometres (nm), of a dust particle.   100

[Total 2 marks]

---

2 The surface area to volume ratio of an object can affect how it behaves.

2.1 Calculate the volume of a cube that has sides of 10 nm long.

Volume = ... nm^3
[1]

2.2 Calculate the total surface area of the same cube (all sides 10 nm long).

Surface area = ... nm^2
[2]

Topic 2 — Bonding, Structure and Properties of Matter

2.3 Use your answers to 2.2 and 2.3 to calculate the surface area to volume ratio
of the cube with 10 nm long sides.

Surface area to volume ratio = ... nm^{-1}
[1]

2.4 Calculate the surface area to volume ratio of a cube with sides of 1 nm and comment on how
changing the side length by a factor of 10 affects the surface area to volume ratio.

Surface area to volume ratio = ...nm^{-1}

Comment: ....

....

[5]

[Total 9 marks]

3 Cube **A** has a surface area to volume ratio of 0.12 nm^{-1}.
Another cube, Cube **B**, has sides that are ten times the length of Cube **A**.   (Grade 7-9)

3.1 What is the surface area to volume ratio of Cube **B**?

Surface area to volume ratio: .... nm^{-1}
[1]

3.2 Material **X** has particles that are a similar size to Cube **A**, and material **Y** has particles
that are a similar size to Cube **B**. Both materials are made from the same compound.
Which material would you need more of in order to catalyse a reaction effectively? Tick **one** box.

Material **X** ☐   Material **Y** ☐

Explain your answer.

....

....

[2]

[Total 3 marks]

Exam Practice Tip

Nanoparticles sound intimidating but don't panic if you come across questions like these in the exam. The maths is
actually pretty simple — it's just calculating areas and volumes and then working out the ratio of one to the other.

 ☐   ☐   ☐

Topic 2 — Bonding, Structure and Properties of Matter

Uses of Nanoparticles

1 Nanoparticles of zinc oxide are used in some sun creams to improve the protection of skin from exposure to sunlight.   Grade 4-6

1.1 State two possible advantages of using nanoparticles of zinc oxide in sun creams.

....

....
                                                                                                                    [2]

1.2 State two possible disadvantages to using nanoparticles of zinc oxide in sun creams.

....

....
                                                                                                                    [2]

[Total 4 marks]

2 The properties of some nanoparticles are listed in **Table 1**.   Grade 6-7

Table 1

Nanoparticle	Properties
Carbon nanotubes	Forms a cage-like structure that can be used to trap small molecules. Light and strong.
Gold	Responds to touch, temperature and humidity. Changes colour in response to the concentration of other compounds in solution.
Silver	Antibacterial

Suggest, with reasoning, which material could be used for the following applications.
You can use each material more than once, and you do not need to use every material.

2.1 Delivering drugs to specific parts of the body.

Material: ....

Reason: ....
                                                                                                                    [2]

2.2 Sterilising water in a water filter.

Material: ....

Reason: ....
                                                                                                                    [2]

2.3 Strengthening light-weight sports equipment, such as tennis racket strings

Material: ....

Reason: ....
                                                                                                                    [2]

[Total 6 marks]

Topic 2 — Bonding, Structure and Properties of Matter

Relative Formula Mass

1 Match up the following formulas with the correct relative formula mass of the substance.

F_2 38

C_2H_6 40

CaO   30

NaOH   56

[Total 2 marks]

2 Magnesium oxide is a salt with the molecular formula MgO.

2.1 Calculate the percentage, by mass, of magnesium ions in magnesium oxide.
Relative atomic masses (A_r): O = 16, Mg = 24

Percentage by mass of magnesium = ... %

[2]

2.2 A chemist is making a mixture that needs to contain 15% magnesium ions by mass.
Calculate the mass of magnesium in 200 g of this mixture.

Mass of magnesium = ... g

[1]

2.3 All the magnesium ions in the mixture come from magnesium oxide.
Using your answers to 2.1 and 2.2, calculate the mass of magnesium oxide needed to provide the correct mass of magnesium ions in 200 g of the mixture.
(If you failed to get an answer to 2.1 or 2.2, you should use the percentage mass of magnesium as 40% and the mass of magnesium ions in the mixture as 20 g. These are **not** the correct values.)

Mass of magnesium oxide = ... g

[1]

[Total 4 marks]

The Mole

The number of atoms, molecules or ions in 1 mole of a substance is given by the Avogadro constant. Tick the box next to the correct value of the Avogadro constant.

6.02×10^{-23} ☐ 0.62×10^{23} ☐ 6.20×10^{-23} ☐ 6.02×10^{23} ☐

1   Carbon dioxide (CO_2) and carbon monoxide (CO) can both be made by reacting carbon and oxygen. Relative atomic masses (A_r): C = 12, O = 16   **Grade 4-6**

1.1   Calculate the relative formula mass (M_r) of carbon dioxide.

M_r of carbon dioxide = ...
[1]

1.2   Calculate the number of moles in 110 g of carbon dioxide.

.... mol
[1]

1.3   Which would weigh more: 1 mole of carbon dioxide or 1 mole of carbon monoxide? Explain your answer using ideas about relative formula mass.

....

....
[2]
[Total 4 marks]

2   Iron and sulfur react together to form iron sulfide in the following reaction: **Grade 7-9**

$$Fe + S \rightarrow FeS$$

2.1   Calculate the mass of 2 moles of sulfur. Relative atomic mass (A_r): S = 32

Mass = ... g
[1]

2.2   Calculate the number of moles in 44 g of iron sulfide.
Relative atomic masses (A_r): Fe = 56, S = 32

.... mol
[2]

2.3   Which is greater, the number of atoms in 3 moles of sulfur or the number of molecules in 2 moles of iron sulfide? Explain your answer.

....

....

....
[2]
[Total 5 marks]

Topic 3 — Quantitative Chemistry

 ☐   ☐   ☐

Conservation of Mass

1 A student burned 12 g of magnesium in oxygen to produce magnesium oxide. *(Grade 4-6)*

1.1 Which of the following is the correctly balanced equation for the reaction between magnesium and oxygen? Tick **one** box.

$Mg + O \rightarrow MgO$ ☐ $2Mg + O_2 \rightarrow 2MgO$ ☐

$Mg + O_2 \rightarrow 2MgO$ ☐ $Mg + O_2 \rightarrow MgO$ ☐

[1]

1.2 The student measured the mass of magnesium oxide produced. The mass was 20 g.
Calculate the mass of oxygen that reacted with the magnesium.

Mass of oxygen = ... g

[2]

[Total 3 marks]

2 This question is about conservation of mass. *(Grade 6-7)*

2.1 Explain what is meant by the term 'conservation of mass'.
Give your answer in terms of the atoms that take part in a chemical reaction.

....

....

....

....

....

[3]

2.2 A student heated zinc powder in air. The equation for the reaction that happened is shown below.

$$2Zn_{(s)} + O_{2(g)} \rightarrow 2ZnO_{(s)}$$

The student weighed the mass of the powder before and after the reaction.
Describe the change that would happen to the mass of the powder during the reaction.
Explain this change using ideas from the particle model.

....

....

....

....

[3]

[Total 6 marks]

Topic 3 — Quantitative Chemistry

3 A student took some sodium carbonate powder and heated it in an open crucible. The equation for the reaction is: $Na_2CO_3 \rightarrow Na_2O + CO_2$

Grade **7-9**

3.1 The student measured that the mass of the powder decreased during the reaction.
She concluded that the measurement must be wrong because mass is conserved during a reaction.
Explain whether the student's measurement or conclusion is likely to be correct.

....

....

....

....

....

[3]

3.2 The student calculated the relative formula masses (M_r) of the reactants and products.
The M_r of sodium carbonate was 106 and the M_r of carbon dioxide was 44.

Use these values to calculate the M_r of sodium oxide.

M_r of sodium oxide = ...
[1]

3.3 In the experiment, the student started with 53 g of sodium carbonate.
Calculate the mass of carbon dioxide that was produced.

Mass of carbon dioxide = ... g
[3]

3.4 Using the law of conservation of mass, calculate the mass of sodium oxide produced.

Mass of sodium oxide = ... g
[1]

[Total 8 marks]

Exam Practice Tip

There's a good chance that if you're given an equation in the exam and asked to do a calculation, you'll probably need to use the molar ratios in the equation. For example, in Q3.3, you need to work out the number of moles in 53 g of Na_2CO_3 and then use the molar ratios in the equation to work out the mass of CO_2.

Topic 3 — Quantitative Chemistry

The Mole and Equations

How many moles of O_2 are shown to be reacting in the equation below? Tick **one** box.

$$4Fe + 3O_2 \rightarrow 2Fe_2O_3$$

6 ☐   5 ☐   2 ☐   3 ☐

1   1 mole of sulfuric acid reacts with 2 moles of sodium hydroxide to form 1 mole of sodium sulfate and 2 moles of water. Which of the following equations shows the reaction? Tick **one** box.

(Grade 4-6)

$2H_2SO_4 + NaOH \rightarrow 2Na_2SO_4 + H_2O$ ☐

$HCl + NaOH \rightarrow NaCl + H_2O$ ☐

$H_2SO_4 + 2NaOH \rightarrow Na_2SO_4 + 2H_2O$ ☐

$Mg + H_2SO_4 \rightarrow MgSO_4 + H_2$ ☐

[Total 1 mark]

2   9.2 g of sodium (Na) was reacted with 7.2 g of water (H_2O) to form sodium hydroxide (NaOH) and hydrogen (H_2). Relative atomic masses (A_r): Na = 23, H = 1, O = 16

(Grade 6-7)

2.1   Calculate the number of moles of sodium that reacted with water.

.... mol
[1]

2.2   Calculate the number of moles of water that reacted with sodium.

.... mol
[2]

2.3   0.4 mol of sodium hydroxide and 0.2 mol of hydrogen were produced in the reaction. Give the balanced symbol equation for the reaction between sodium and water.

Balanced symbol equation: .... + ... $\rightarrow$ .... + ...
[3]

[Total 6 marks]

3 Methane (CH_4) reacts with oxygen (O_2) to form carbon dioxide and water.
Relative formula masses (M_r): $CH_4 = 16$, $O_2 = 32$, $CO_2 = 44$, $H_2O = 18$

(Grade 7-9)

3.1 A student carried out this reaction, reacting 8 g of methane with 32 g of oxygen, producing 22 g of carbon dioxide and 18 g of water.

Use these masses to work out the balanced equation for the reaction between methane and oxygen. Show your working.

Balanced symbol equation: .... + ... → .... + ...

[3]

3.2 The student repeated the reaction with 48 g of oxygen.
Calculate the number of moles of carbon dioxide that were produced.

.... mol

[3]

3.3 Another student carried out the reaction using 4 mol of methane.
Calculate the number of moles of water produced in this reaction.

.... mol

[1]

3.4 Calculate the mass of water produced in the reaction using 4 mol of methane.

Mass of water = ... g

[1]

[Total 8 marks]

Exam Practice Tip

If you're given the mass of a substance then you can use the formula linking mass, M_r and moles to calculate the number of moles. But remember, you can rearrange the formula too — so if you know how many moles of something you have, you can work out its mass, or if you know the mass and the number of moles you can work out its M_r. Cool eh?

Topic 3 — Quantitative Chemistry

Limiting Reactants

1   A student reacted excess calcium with hydrochloric acid.   *(Grade 4-6)*

1.1   Explain why calcium was added in excess.

....

....
   [1]

1.2   In this reaction, the hydrochloric acid is called the 'limiting reactant'.
Explain what this term means.

....

....
   [2]

[Total 3 marks]

2   A student reacted copper oxide with sulfuric acid to make copper sulfate and water.   *(Grade 7-9)*

$$CuO + H_2SO_4 \rightarrow CuSO_4 + H_2O$$

2.1   The student used 0.50 mol of sulfuric acid and an excess of copper oxide.
What mass of copper sulfate would be produced by the reaction?

Relative atomic masses (A_r): Cu = 63.5, S = 32, O = 16, H = 1

Mass of copper sulfate = ... g
   [3]

2.2   The student decides to double the quantity of the sulfuric acid and use an excess of copper oxide.
Describe how this would affect the yield of the copper sulfate. Explain your answer.

....

....
   [2]

2.3   The student found that only 0.4 mol of copper oxide was available
to react with the doubled quantity of sulfuric acid in question 2.2.
Explain the effect this would have on the amount of product obtained.

....

....

....

....
   [3]

[Total 8 marks]

Topic 3 — Quantitative Chemistry

Gases and Solutions

1   28 g of calcium chloride was dissolved in 0.4 dm³ of water. $\overset{\text{Grade}}{\text{4-6}}$

1.1   Calculate the concentration of the solution and give the units.

Concentration = ... Units = ...
[2]

1.2   Explain the term 'concentration of a solution'.

....

....
[1]

[Total 3 marks]

2   Carbon dioxide is a gas with a relative formula mass of 44. $\overset{\text{Grade}}{\text{6-7}}$

2.1   Calculate the mass of 36 dm³ of CO_2, at r.t.p.

Mass = ... g
[2]

2.2   Compare the volumes of 1 mole of CO_2 and 1 mole of O_2, at r.t.p. Explain your answer.

....

....
[2]

[Total 4 marks]

3   A student burned methane with 16 g of oxygen to produce water and carbon dioxide.
Relative atomic masses (A_r): C = 12, H = 1, O = 16 $\overset{\text{Grade}}{\text{6-7}}$

$$CH_{4(g)} + 2O_{2(g)} \rightarrow CO_{2(g)} + 2H_2O_{(l)}$$

3.1   Calculate the volume of oxygen used in the reaction, at r.t.p.

Volume = ... dm³
[3]

3.2   Using the reaction equation and the amount of oxygen burned,
calculate the volume of CO_2 produced at r.t.p.

Volume = ...dm³
[2]

[Total 5 marks]

Topic 3 — Quantitative Chemistry

4 Carbon dioxide can be produced by reacting oxygen with carbon monoxide.
Relative atomic masses (A_r): C = 12, O = 16

$$2CO_{(g)} + O_{2(g)} \rightarrow 2CO_{2(g)}$$

4.1 A student reacted 48 dm³ of carbon monoxide with oxygen.
Calculate the volume of oxygen that reacted, at r.t.p.

Volume of oxygen =.............. dm³
[1]

4.2 Another student reacted 28 g of carbon monoxide with oxygen.
Calculate the volume of oxygen involved in the reaction.

Volume of oxygen =.............. dm³
[4]

4.3 Calculate the volume of carbon dioxide produced in the reaction in 4.2.

Volume of carbon dioxide =.............. dm³
[1]

[Total 6 marks]

5 A student made a solution of sodium carbonate (Na_2CO_3) at a concentration of 0.50 mol/dm³.

5.1 Calculate the number of moles of sodium carbonate in 0.50 dm³ of the solution.

.... mol
[1]

5.2 Calculate the mass of sodium carbonate needed to make 0.50 dm³ of the solution.
Relative atomic masses (A_r): Na = 23, C = 12, O = 16

Mass of sodium carbonate = ... g
[2]

[Total 3 marks]

Exam Practice Tip

The rule that one mole of a gas occupies 24 dm³ is only true at 20 °C and 1 atm of pressure, otherwise known as room temperature and pressure, or 'r.t.p.' If a question says 'at r.t.p.', it's a bit of a clue that you need to use this rule.

Topic 3 — Quantitative Chemistry

Concentration Calculations

1   25.0 cm³ of 18.25 g/dm³ hydrochloric acid (HCl) is needed to neutralise 50.0 cm³ of sodium hydroxide solution (NaOH). The equation for the reaction is:

$$HCl + NaOH \rightarrow NaCl + H_2O$$

1.1 Calculate the concentration of the hydrochloric acid in mol/dm³.
Relative atomic masses (A_r): H = 1, Cl = 35.5

Concentration of hydrochloric acid = ... mol/dm³
[3]

1.2 Calculate the number of moles of hydrochloric acid that reacted with the sodium hydroxide.

Moles of hydrochloric acid = ... mol
[3]

1.3 Calculate the number of moles of sodium hydroxide that reacted with the hydrochloric acid.

Moles of sodium hydroxide = ... mol
[1]

1.4 Calculate the concentration of the sodium hydroxide solution, in mol/dm³.

Concentration of sodium hydroxide = ... mol/dm³
[3]

1.5 Convert the concentration of sodium hydroxide to g/dm³.
Relative atomic masses (A_r): Na = 23, O = 16, H = 1

Concentration of sodium hydroxide = ... g/dm³
[3]

[Total 13 marks]

Topic 3 — Quantitative Chemistry

2 A student had a solution of sodium carbonate (Na_2CO_3) with an unknown concentration. The student also had some hydrochloric acid (HCl) at a concentration of 1.00 mol/dm³. The word equation for the neutralisation reaction between them is:

sodium carbonate + hydrochloric acid → sodium chloride + water + carbon dioxide

2.1 Give the balanced symbol equation for this reaction.

....

[2]

2.2 The student carried out an experiment to find out how much hydrochloric acid was needed to neutralise 25.0 cm³ of the sodium carbonate solution. They did the experiment 5 times and their results are shown in **Table 1**.

Calculate the mean volume of hydrochloric acid that was needed.
Ignore any anomalous results (ones that are not within 0.10 cm³ of each other.)

Table 1

	Experiment Number				
	1	2	3	4	5
Volume of 1.00 mol/dm³ HCl (cm³)	12.95	12.50	12.25	12.55	12.45

Mean volume = ... cm³

[2]

2.3 Using your answer to 2.2, calculate the concentration of the 25.0 cm³ solution of sodium carbonate, in mol/dm³.

Concentration of sodium carbonate = ... mol/dm³

[6]

[Total 10 marks]

Exam Practice Tip

When you're working out the mean from a set of results, you should ignore any anomalous results. So if you get a question in your exam asking you to work out a mean, make sure you check for anomalous results before you start plugging numbers into your calculator. The question might help you out by reminding you to ignore anomalous results, but it might not, so make sure you're on the ball with it — don't get caught out.

Topic 3 — Quantitative Chemistry

Atom Economy

1 Ethene is often produced by heating ethanol in the presence of a catalyst. The equation for this reaction is: $C_2H_5OH \rightarrow C_2H_4 + H_2O$. Atomic masses ($A_r$): C = 12, H = 1, O = 16.

Grade 4-6

1.1 Give two reason why a high atom economy is desirable in industry.

....

....

[2]

1.2 Calculate the relative formula mass of ethanol.

M_r of ethanol = ...

[1]

1.3 Calculate the relative formula mass of ethene.

M_r of ethene = ...

[1]

1.4 Calculate the atom economy of the reaction.

Atom economy = ... %

[2]

[Total 6 marks]

2 A student wanted to produce a sample of hydrogen gas by reacting either magnesium (Mg) or zinc (Zn) with hydrochloric acid (HCl). The equations for these reactions are: $Mg + 2HCl \rightarrow MgCl_2 + H_2$ and $Zn + 2HCl \rightarrow ZnCl_2 + H_2$.

Grade 7-9

Determine which reaction is more economical by calculating and comparing the atom economy of both reactions.
Relative formula masses (M_r): HCl = 36.5, H_2 = 2
Relative atomic masses (A_r): Mg = 24, Zn = 65

Atom economy of reaction using magnesium = ... %

Atom economy of reaction using zinc = ... %

More economical reaction: ....

[Total 5 marks]

Percentage Yield

1 A student produced magnesium oxide by burning magnesium in air. The student calculated the theoretical yield of magnesium oxide to be 2.4 g. The actual yield of magnesium oxide was 1.8 g.

1.1 Calculate the percentage yield.

Percentage yield = ... %

[2]

1.2 The student's method involved heating a piece of magnesium of known mass in a crucible until the reaction appeared to have finished. The product was then tipped onto a balance and weighed. Suggest one way that this method may have decreased the percentage yield.

....

[1]

[Total 3 marks]

2 Ammonia is produced in the Haber process by reacting nitrogen gas with hydrogen gas. The equation for this reaction is: $N_2 + 3H_2 \rightleftharpoons 2NH_3$

2.1 A factory wanted to use 14 kg of nitrogen gas to produce ammonia. Calculate the theoretical yield of ammonia in this reaction. Atomic masses (A_r): N = 14, H = 1

Theoretical yield = ... kg

[4]

2.2 After the reaction, the factory had produced 4.5 kg of ammonia. Calculate the percentage yield for the reaction.

Percentage yield = ... %

[2]

2.3 Give two reasons why the actual yield of ammonia was lower than the theoretical yield.

....

....

[2]

2.4 Give two reasons why it is desirable for a factory to obtain as high a percentage yield as possible.

....

....

[2]

[Total 10 marks]

Topic 3 — Quantitative Chemistry

Topic 4 — Chemical Changes

Acids and Bases

Fill in the gaps for the following paragraph on techniques to measure pH.

red
neutral   less
purple
green
more

Universal indicator will turn .... in strongly acidic
solutions and .... in strongly alkaline solutions.
In a .... solution, Universal indicator will be green.
A pH probe attached to a pH meter is .... accurate
than Universal indicator as it displays a numerical value for pH.

1   This question is about acids and bases. **Table 1** shows some everyday substances.   **Grade 4-6**

Table 1

Substance	pH
Beer	4
Bicarbonate of soda	9
Milk	7

1.1 Write the name of the substance in **Table 1** that is an acid.

....

[1]

1.2 What colour would you expect to see if Universal indicator was added
to bicarbonate of soda solution?

....

[1]

1.3 Which ion is produced by an acid in aqueous solution?
Tick **one** box.

☐ Cl⁻   ☐ H⁺   ☐ OH⁻   ☐ OH⁺

[1]

1.4 State the range of the pH scale.

....

[2]

[Total 5 marks]

2   Acids and alkalis react together in neutralisation reactions.   **Grade 4-6**

2.1 Write the general word equation for a neutralisation reaction between an acid and an alkali.

....

[1]

2.2 In terms of hydrogen ions and hydroxide ions, write an ionic equation for a neutralisation reaction.

....

[1]

[Total 2 marks]

Titrations

Warm-Up

Label the diagram using the labels below. The first one has been done for you.

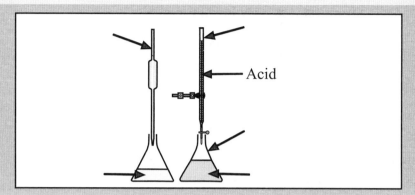

Solution containing an alkali and an indicator

Alkali   ~~Acid~~

Burette

Pipette

Conical flask

— Acid

1 A student wants to find out the concentration of a solution of alkali by titrating it with an acid. Grade 4-6

1.1 Identify which of the following statements is false. Tick **one** box only.

Universal indicator is the most suitable indicator for
use in titrations. ☐

Titrations can be used to determine the concentration
of an unknown solution. ☐

Concentration can be measured in g/dm^3 or mol/dm^3. ☐

An indicator is usually used to identify the point
of neutralisation in an acid-base titration. ☐

[1]

1.2 Explain why a burette is useful for identifying the end-point of a titration.

....

....

....
[2]

1.3 Here is the method the student uses for the titration:

1. Add the acid to the alkali from the burette a
   little at a time, regularly swirling the conical flask.
2. Record the volume of acid required to just make the indicator change colour.
3. Use this volume to calculate the concentration of the alkali.

Explain how the method should be changed to increase the accuracy of the results.

....

....
[2]

[Total 5 marks]

Topic 4 — Chemical Changes

Strong Acids and Weak Acids

1 Below are some acids listed in order of strength.

Strongest ↑ Hydrochloric Acid

Nitric Acid

Citric Acid

Ethanoic Acid

Weakest ↓ Carbonic Acid

1.1 Hydrochloric acid is a strong acid while carbonic acid is a weak acid.
Explain the difference between a strong acid and a weak acid.

....

....

[2]

1.2 Describe and explain the difference you would expect to find in the pH value of a 1 mol/dm³
solution of nitric acid and a 1 mol/dm³ solution of ethanoic acid.

....

....

....

[3]

1.3 A 0.01 mol/dm³ solution of hydrochloric acid has a pH of 2.
Predict the pH of a 0.001 mol/dm³ solution of hydrochloric acid.

....

[1]

1.4 A solution of citric acid in a beaker has a pH of 3. Which of the following would increase the pH?
Tick **three** boxes that apply .

Adding citric acid with a higher concentration to the beaker. ☐

Adding water to the beaker. ☐

Adding ethanoic acid to the beaker at the same concentration as the citric acid. ☐

Adding hydrochloric acid to the beaker at the same concentration as the citric acid. ☐

Changing the citric acid to nitric acid of the same concentration. ☐

Changing the citric acid to carbonic acid of the same concentration. ☐

[3]

[Total 9 marks]

Exam Practice Tip

Make sure you understand what pH is and how it relates to the strength of an acid. It will also be jolly useful to
understand how the concentration of H⁺ changes with pH. Remember, don't confuse strong acids with concentrated ones.

Topic 4 — Chemical Changes

Reactions of Acids

1   Acids can react with bases and alkalis.   Grade 4-6

1.1   What type of reaction occurs when an acid reacts with an alkali?
Tick **one** box.

☐ Oxidation ☐ Decomposition ☐ Neutralisation ☐ Precipitation

[1]

1.2   A student adds 2 spatulas of zinc carbonate into a beaker of dilute hydrochloric acid.
Draw **one** line to connect the observation you would expect during this reaction
with the explanation for that observation.

Observation

| Fizzing |

| A colour change |

| A white solid forming |

| No change |

Explanation

| Carbon dioxide is produced |

| Hydrogen is produced |

| A salt is made |

| The zinc dissolves |

[1]

[Total 2 marks]

2   Acids can react with metal oxides and metal hydroxides to give a metal salt and water.   Grade 6-7

2.1   Write a word equation for the reaction of sulfuric acid and lithium hydroxide.

....

[2]

2.2   Complete and balance the symbol equation for the reaction of sulfuric acid and lithium hydroxide.

H_2SO_4 + ... LiOH → .... + ... H_2O

[2]

2.3   Acids can also react with metal carbonates. Compare the products of the reaction between
sulfuric acid and lithium carbonate with the products of the reaction between sulfuric acid and
lithium hydroxide.

....

....

[2]

[Total 5 marks]

3 Soluble metal salts can be made from the reactions of acids and metal oxides.

3.1 Zinc chloride can be made from the reaction of zinc oxide and hydrochloric acid. Describe a laboratory method to produce pure crystals of zinc chloride using this reaction.

....

....

....

....

....

....

....

[4]

3.2 Suggest an alternative to zinc oxide that would also react with hydrochloric acid to form the desired product.

....

[1]

[Total 5 marks]

4* A student is given three different unlabelled solutions. One contains sodium hydroxide, one contains sodium carbonate and one contains a sodium salt. Use your knowledge of chemical reactions to describe experiments that the student could do to decide which solution is which. Explain how your experiments would allow the student to identify the solutions. Clearly describe the reactants and products of any reactions you include.

....

....

....

....

....

....

....

....

....

....

....

[Total 6 marks]

Topic 4 — Chemical Changes

The Reactivity Series

1 Metals can be placed in order of reactivity based on how vigorously they react with acids. *(Grade 4-6)*

1.1 Write a word equation for the reaction of magnesium with hydrochloric acid.

....
[1]

1.2 What does the reaction of magnesium with hydrochloric acid produce? Tick **one** box.

Positive chloride ions ☐

Positive hydrogen ions ☐

Positive magnesium ions ☐

Negative magnesium ions ☐

[1]

1.3 Explain why iron reacts less vigorously with hydrochloric acid than magnesium does.

....
[1]

1.4 Name **one** metal that would react more vigorously with hydrochloric acid than magnesium does.

....
[1]

[Total 4 marks]

2 Some metals can react with water. *(Grade 4-6)*

2.1 Write the general word equation for the reaction of a metal and water.

....
[1]

2.2 Complete the symbol equation below, for the reaction of calcium and water. Include state symbols.

$$Ca_{(s)} + 2H_2O_{(l)} \rightarrow \text{.................} + \text{.................}$$
[2]

2.3 Suggest a metal which will react more vigorously with water than calcium. Explain your answer.

....

....
[2]

2.4 Put the metals sodium, zinc and potassium in order, based on how vigorously you think they would react with water.

Most vigorous .... Least vigorous
[1]

[Total 6 marks]

Topic 4 — Chemical Changes

3   A student investigated the reactions of some metals and found the results shown in **Table 1**.

Table 1

Reaction	Observation
Lithium + water	Very vigorous reaction with fizzing, lithium disappears
Calcium + water	Fizzing, calcium disappears
Magnesium + water	No fizzing, a few bubbles on the magnesium
Copper + water	No fizzing, no change to copper
Iron + water	No fizzing, no change to iron
Lithium + dilute acid	Very vigorous reaction with fizzing, lithium disappears
Magnesium + dilute acid	Fizzing, magnesium disappears
Zinc + dilute acid	Fizzing, zinc disappears
Copper + dilute acid	No fizzing, no change to copper

3.1   Use the results in **Table 1**, along with your knowledge of the general reaction between an acid and a metal, to explain whether lithium or magnesium forms positive ions more easily.

....

....

....

[3]

3.2   Predict what the student would have seen if they had added sodium to water.

....

....

[2]

3.3   Put the metals calcium, copper and lithium in order from most reactive to least reactive.

....

[1]

3.4   Explain why it would be difficult to decide the order of reactivity of magnesium and zinc using these experiments. Suggest an experiment that could be used to decide which is most reactive.

....

....

[2]

[Total 8 marks]

Exam Practice Tip

Learning the order of the reactivity series could be really useful when it comes to answering questions in the exams. Try learning this mnemonic to help you remember... Papa Smurf Likes Calling My Clarinet Zany — Isn't He Cute. (You don't have to use my Booker prize winning concoction, though. You could also make up your own.)

Topic 4 — Chemical Changes

Separating Metals from Metal Oxides

1 Most metals are found as compounds in ores in the earth. Some of the metals can be extracted from their ores by reduction with carbon.

Grade 4-6

1.1 Name a metal that can be found in the ground as an element.

....

[1]

1.2 Why are most metals found in the earth as compounds?

....

[1]

1.3 Define reduction in terms of the loss and gain of oxygen.

....

[1]

1.4 Explain why magnesium **cannot** be extracted from magnesium oxide by reduction with carbon.

....

....

[1]

[Total 4 marks]

2 Iron can be extracted by heating iron oxide, Fe_2O_3, with carbon. The reaction releases large amounts of heat so there is no need to continuously heat the reaction.

Grade 6-7

2.1 Write a symbol equation for a reaction between carbon and iron(III) oxide, Fe_2O_3.

....

[2]

2.2 For the reaction of iron(III) oxide and carbon to form iron, identify whether carbon has been reduced or oxidised. Explain your answer in terms of the transfer of oxygen.

....

....

[2]

2.3 Magnesium is extracted from magnesium oxide by reaction with silicon.
A temperature of 1200 °C is required together with a reduced pressure.

Suggest why magnesium extraction is very costly when compared with iron extraction.

....

....

....

[2]

[Total 6 marks]

Redox Reactions

1 In a metal displacement reaction the least reactive metal is reduced. **Grade 6-7**

1.1 Define reduction in terms of electron transfer.

....
[1]

1.2 Which of the following reactions shows zinc being reduced?
Tick **one** box.

Copper nitrate + zinc → copper + zinc nitrate ☐

Zinc + oxygen → zinc oxide ☐

Zinc chloride + sodium → zinc + sodium chloride ☐

Zinc + hydrochloric acid → zinc chloride + hydrogen ☐
[1]

1.3 The equation shows the reaction of zinc with hydrochloric acid.

$$Zn_{(s)} + 2HCl_{(aq)} \rightarrow ZnCl_{2(aq)} + H_{2(g)}$$

Does hydrogen lose or gain electrons in this reaction?

....
[1]

1.4 What happens to chlorine in the reaction in 1.3?
Tick **one** box.

Chlorine is oxidised ☐

Chlorine is released ☐

Chlorine is reduced ☐

Chlorine is neither oxidised nor reduced ☐
[1]

[Total 4 marks]

2 A student reacts magnesium with an aqueous solution of iron chloride to produce magnesium chloride and iron. **Grade 7-9**
$$Mg_{(s)} + FeCl_{2(aq)} \rightarrow MgCl_{2(aq)} + Fe_{(s)}$$

2.1 Write an ionic equation for this reaction.

....
[1]

2.2 A student repeats the experiment with copper instead of magnesium.
State whether a reaction would still occur. Explain your answer.

....

....
[2]

[Total 3 marks]

Topic 4 — Chemical Changes

Electrolysis

Place the labels on the correct label lines to identify the parts of an electrochemical cell.

Anode   Electrolyte

D.C. power supply

Cathode

Anions

Cations

1   Lead bromide can be electrolysed, using molten lead bromide as the electrolyte. (Grade 4-6)

1.1 What is an electrolyte?

....

[1]

1.2 Write the word equation for the electrolysis of lead bromide.

....

[1]

1.3 Explain why lead ions move towards the cathode and not the anode.

....

....

[2]

1.4 What ions move towards the anode? Give the chemical formula and charge of the ion.

....

[1]

1.5 Is the reaction at the anode oxidation or reduction?

....

[1]

1.6 Why does the lead bromide need to be molten? Tick **one** box.

So the ions can move to the electrodes ☐

So the electrons can be conducted through the substance ☐

So the electrodes don't corrode ☐

So there is enough heat for the reaction to occur ☐

[1]

[Total 7 marks]

2   **Figure 1** shows the extraction of aluminium. Aluminium oxide is mixed with cryolite. This mixture is then melted and electrolysed. Metallic aluminium is made at the cathode.

Figure 1

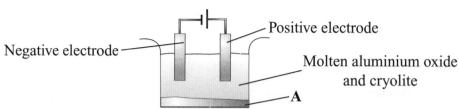

2.1 What is the liquid labelled **A**?

....

[1]

2.2 What is the purpose of mixing the aluminium oxide with cryolite?

....

[1]

2.3 Why do the graphite electrodes need to be replaced regularly?

....

....

[2]

[Total 4 marks]

3   Aqueous iron chloride solution can be electrolysed using inert electrodes.   Grade 6-7

3.1 Write the names of the ions present in iron chloride solution.

....

[2]

3.2 Draw **one** line to connect the correct products at each electrode when iron chloride is electrolysed.

At the cathode	**At the anode**
Iron is discharged	Iron is discharged
Hydrogen is discharged	Oxygen is discharged
Chlorine is discharged	Chlorine is discharged

[1]

3.3 What is discharged at the anode when iron sulfate solution is electrolysed with inert electrodes?

....

[1]

3.4 Iron can be extracted from iron solutions by electrolysis but this is not the usual method. Why is electrolysis not the usual method of extracting iron?

....

....

[2]

[Total 6 marks]

Topic 4 — Chemical Changes

4   A student investigated the products of electrolysis of a variety of aqueous solutions using inert electrodes. **PRACTICAL**

4.1 Draw a labelled diagram of suitable apparatus that could be used for these experiments.

[4]

4.2 Complete **Table 1** by predicting the products at the anode and cathode for each of the solutions.

Table 1

Solution	Product at cathode	Product at anode
$CuCl_2$		
KBr		
H_2SO_4		

[6]

4.3 When potassium nitrate solution is electrolysed neither potassium nor nitrogen are discharged. Explain why and state what is produced instead.

....

....

....

[4]

4.4 Write two half equations for the reaction that occurs when water is electrolysed.

Cathode: ....

Anode: ....

[2]

[Total 16 marks]

Exam Practice Tip

Electrolysis can be a hard subject to get your head around, and adding the electrolysis of aqueous solutions in to the mix doesn't make it any easier. But remember, in aqueous solution, different ions can be discharged depending on their reactivity. Make sure you know the different ions that can be removed from solution, and in what situations that will happen — it really isn't too complicated once you know what you are doing, but you do need to learn the rules.

                      Topic 4 — Chemical Changes

Exothermic and Endothermic Reactions

1   Which one of the following statements about exothermic and endothermic reactions is correct? Tick **one** box.   *(Grade 4-6)*

In an exothermic reaction, energy is transferred from the surroundings so the temperature of the surroundings goes down. ☐

In an endothermic reaction, energy is transferred from the surroundings so the temperature of the surroundings goes down. ☐

In an exothermic reaction, energy is transferred from the surroundings so the temperature of the surroundings goes up. ☐

In an endothermic reaction, energy is transferred from the surroundings so the temperature of the surroundings goes up. ☐

[Total 1 mark]

2   During a reaction between solutions of citric acid and sodium hydrogen carbonate, the temperature of the reaction mixture fell from 18 °C to 4 °C.   *(Grade 4-6)*

2.1   Is this reaction exothermic or endothermic?

....

[1]

2.2   Complete the reaction profile for this reaction to show how the energy changes as the reactants form the products. Mark the overall energy change on the diagram.

[2]

2.3   Where is the energy being transferred from in this type of reaction?

....

[1]

2.4   What happens to the amount of energy in the universe after the reaction?

....

[1]

2.5   Give a practical use of this type of reaction.

....

[1]

[Total 6 marks]

3 A company is trying to find a reaction with a low activation energy to use in a hand warmer. The reaction profiles for the reactions being investigated are shown in **Figure 1**.

Grade 6-7

Figure 1

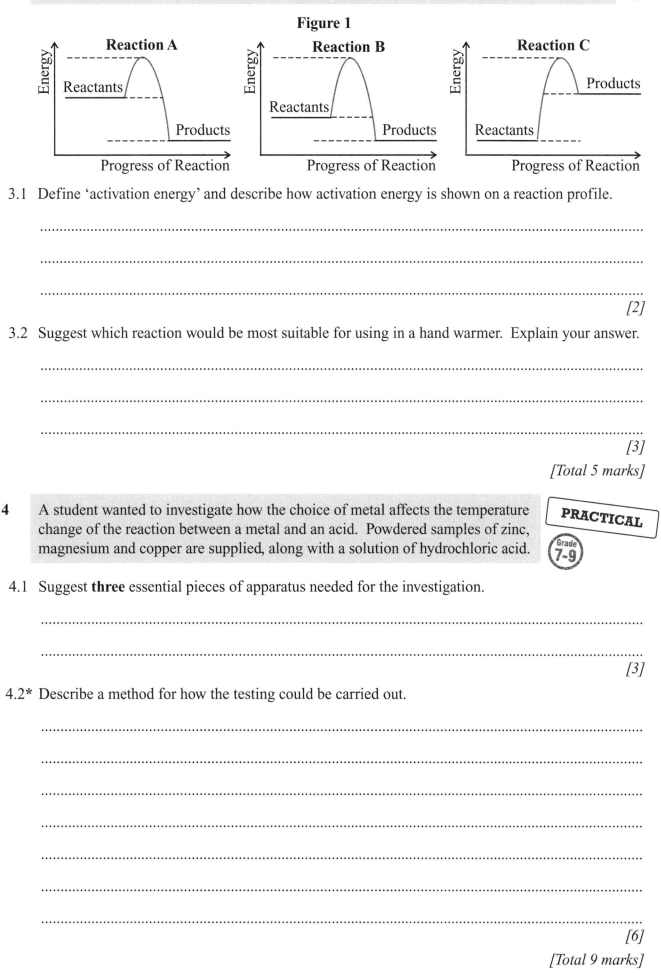

3.1 Define 'activation energy' and describe how activation energy is shown on a reaction profile.

....

....

....

[2]

3.2 Suggest which reaction would be most suitable for using in a hand warmer. Explain your answer.

....

....

....

[3]

[Total 5 marks]

4 A student wanted to investigate how the choice of metal affects the temperature change of the reaction between a metal and an acid. Powdered samples of zinc, magnesium and copper are supplied, along with a solution of hydrochloric acid.

PRACTICAL

Grade 7-9

4.1 Suggest **three** essential pieces of apparatus needed for the investigation.

....

....

[3]

4.2* Describe a method for how the testing could be carried out.

....

....

....

....

....

....

....

[6]

[Total 9 marks]

Topic 5 — Energy Changes

Bond Energies

1 Methane and chlorine can react together as shown in **Figure 1**.
The bond energies of the substances in this reaction are shown in **Table 1**.

Figure 1

H—C—H + Cl—Cl ⟶ H—C—Cl + H—Cl

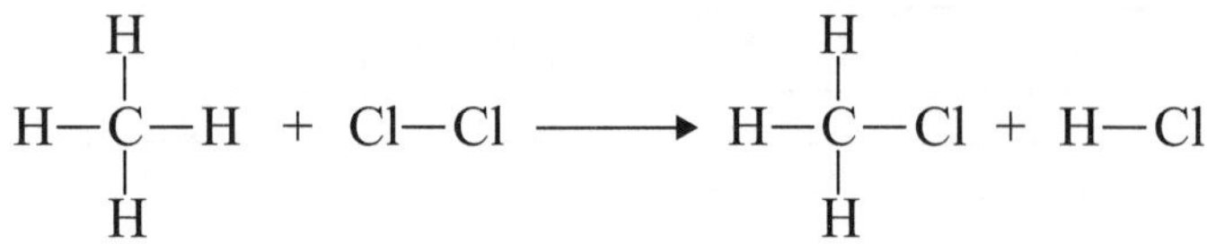

Table 1

Bond	Energy (kJ/mol)
C–H	413
Cl–Cl	243
C–Cl	346
H–Cl	432

1.1 Using the data in **Table 1**, calculate the energy change for the reaction.

Energy change of the reaction = ... kJ/mol
[3]

1.2 Explain, in terms of bond energies, whether the reaction is endothermic or exothermic.

....

....
[2]

[Total 5 marks]

2 Hydrogen and fluorine react together in the following way: $H_2 + F_2 \rightarrow 2HF$.
The overall energy change of the reaction is –542 kJ/mol.
The H–H bond energy is 436 kJ/mol and the F–F bond energy is 158 kJ/mol.

Calculate the energy of the H–F bond.

H–F bond energy = ... kJ/mol
[Total 3 marks]

Exam Practice Tip

The maths for bond energy calculations isn't too tricky — it's usually just simple addition, subtraction and multiplication, and you might need to rearrange an equation. But it's really, really easy to make a mistake — so take your time, work through the question carefully, and, if you've got time at the end of the exam, go back and check your answers.

Cells, Batteries and Fuel Cells

For each of the following statements, circle whether they are **true** or **false**.

1) A hydrogen fuel cell can be recharged.   True   or   False

2) The by-products from hydrogen fuel cells are water and carbon dioxide.   True   or   False

3) In a hydrogen fuel cell, the hydrogen fuel is oxidised.   True   or   False

1   Draw lines from the chemicals below to the best description of the role they play in a hydrogen fuel cell. Draw **three** lines.

hydrogen		waste product
potassium hydroxide		electrolyte
water		fuel

[Total 2 marks]

2   A student, investigating chemical cells, set up the following two cells.

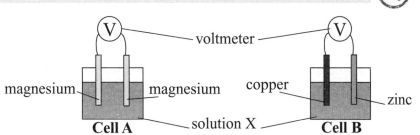

2.1   Given that the greater the difference in reactivity of the metal electrodes, the greater the voltage, predict which cell would produce the higher voltage. Give a reason for your answer.

....

....

[2]

2.2   The student connected two of Cell **B** together in series.
Give the name of the set-up produced, and state the effect it will have on the voltage.

....

....

[2]

2.3   The student uses Cell **B** to power a light bulb. Explain why the light bulb eventually goes out.

....

....

[2]

2.4   Cell **B** is rechargeable. What does this tell you about the type of reaction in the cell?

....

[1]

[Total 7 marks]

Topic 5 — Energy Changes

3 Hydrogen is used in fuel cells, as shown in **Figure 2**. The hydrogen gas reacts at the anode.

Figure 2

hydrogen in ➡ ⟵ oxygen in

anode (–ve electrode)
solution of phosphoric acid
waste product out
cathode (+ve electrode)

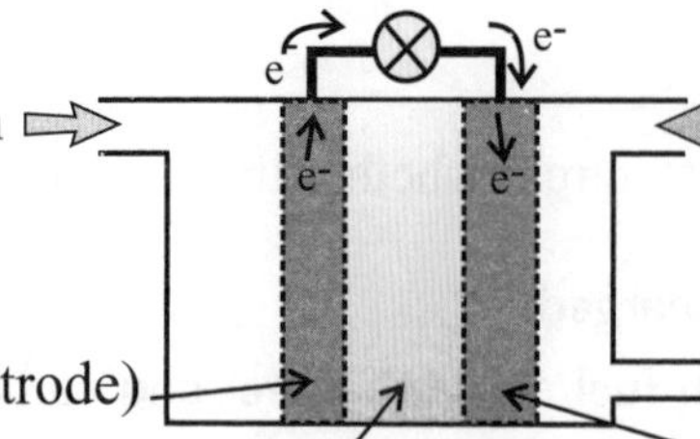

3.1 Write a half equation for the reaction at the anode. What type of reaction is this?

Equation: ....

Type of reaction: ....

[2]

3.2 What is the waste product produced by the hydrogen fuel cell? ....

[1]

3.3 Suggest **one** advantage of using hydrogen fuel cells rather than rechargeable batteries in cars.

....

[1]

[Total 4 marks]

4 A student investigated the reactivity of four unknown metals (labelled A to D) by using them as electrodes in simple cells as shown in **Figure 3**. The results are shown in **Table 1**.

Figure 3

metal 1   metal 2

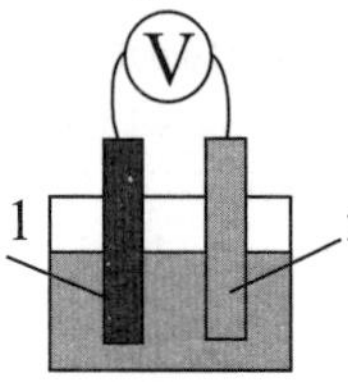

Table 1

Metal 1	silver	silver	silver	silver
Metal 2	A	B	C	D
Voltage (V)	1.56	0.34	3.18	1.05

4.1 In each case, silver was the least reactive metal. Use the data in **Table 1** to place the four metals, A, B, C and D in order of reactivity, starting with the most reactive. Explain your answer.

Order: ....

Explanation: ....

....

....

[4]

4.2 Apart from the metals, what else could be changed that may affect the voltage?

....

[1]

[Total 5 marks]

Topic 5 — Energy Changes

Rates of Reaction

1   This question is about the rate of a chemical reaction between two reactants, one of which is in solution, and one of which is a solid.

Grade 4-6

1.1   Which of the following changes would **not** cause the rate of the chemical reaction to increase? Tick **one** box.

Increasing the concentration of the solutions. ☐

Heating the reaction mixture to a higher temperature. ☐

Using a larger volume of the solution, but keeping the concentration the same. ☐

Grinding the solid reactant so that it forms a fine powder. ☐

[1]

1.2   What is the name given to the minimum amount of energy which particles must have if they are to react when they collide?

....

[1]

1.3   Explain why adding a catalyst to the reaction mixture, without changing any other condition, can cause the rate of the reaction to increase.

....

[1]

[Total 3 marks]

2   **Figure 1** shows how the mass of gas lost from a reaction vessel changes over time, for the same reaction under different conditions.

Grade 6-7

Figure 1

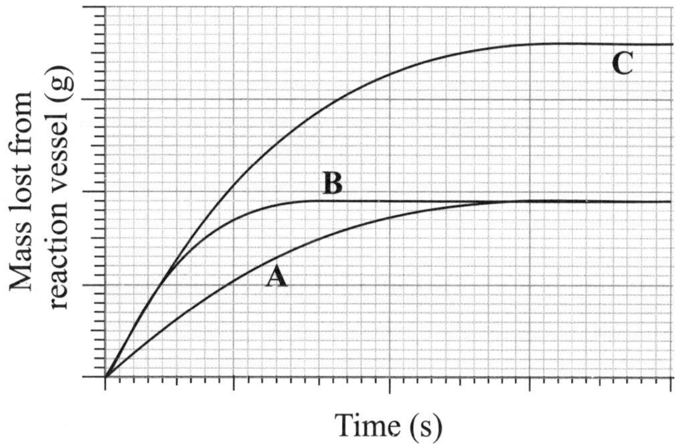

State which of the reactions, **A**, **B** or **C**:

Produced the most product: ....

Finished first: ....

Started at the slowest rate: ....

[Total 3 marks]

3 This question is about the rate of the reaction between magnesium and hydrochloric acid. The chemical equation for the reaction is:

$$Mg_{(s)} + 2HCl_{(aq)} \rightarrow MgCl_{2(aq)} + H_{2(g)}$$

3.1 The graph in **Figure 2** shows how the volume of hydrogen produced changes over the course of the reaction when a small lump of magnesium is added to excess hydrochloric acid.

On the same axes, sketch a curve to show how the volume of hydrogen produced would change over time if an identical piece of magnesium was added to excess hydrochloric acid with a higher concentration.

Figure 2

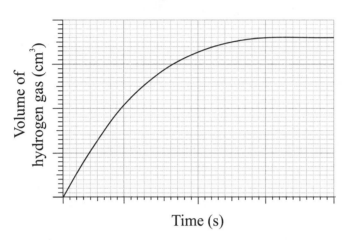

[3]

3.2 State **two** properties of the collisions between particles that affect the rate of a reaction.

....

....

[2]

3.3 Use collision theory to explain why increasing the concentration of a reactant increases the rate of the reaction.

....

....

[2]

3.4 How would you expect the reaction rate to change if the magnesium was cut into smaller pieces?

....

[1]

3.5 Explain why cutting the magnesium into smaller pieces affects the rate of this reaction.

....

....

[2]

3.6 State **one** change that could be made to change the rate of the reaction, other than changing the concentration and the size of the magnesium pieces.

....

[1]

[Total 11 marks]

Topic 6 — The Rate and Extent of Chemical Change

4 A student carries out a reaction between hydrogen gas and iodine gas to form hydrogen iodide (HI).

4.1 The student wants to increase the rate of the reaction, and decides to try two techniques:

    1. Increasing the volume of the reaction vessel.
    2. Increasing the temperature at which the reaction is carried out.

Explain, using collision theory, whether or not these changes will increase the rate of the reaction.

....

....

....

....

....

....

[5]

4.2 When cerium oxide is mixed with the hydrogen and iodine gases, the rate of the reaction increases. What does this suggest about cerium oxide?

....

[1]

4.3 State whether the reaction equation will change if cerium oxide is present in the reaction vessel. Explain your answer.

....

....

[2]

4.4 Sketch and label **two** reaction profiles on the axes below to show the difference between the reaction of hydrogen and iodine with and without cerium oxide. The energy of the products is lower than the energy of the reactants.

[3]

[Total 12 marks]

Exam Practice Tip

The key to explaining questions about reaction rates is usually collision theory. So if these questions are making you bang your head against your desk, have a read up on how collisions affect the rate of a reaction, and on how different reaction conditions influence these collisions. You'll thank me for it later, I can promise you.

Topic 6 — The Rate and Extent of Chemical Change

PRACTICAL | Measuring Rates of Reaction

Warm-Up

Use the words on the left below to fill in the blanks of the passage on the right below.
No word is used more than once, but you may not need to use all the words.

products   mass

tangent

gradient   temperature

reactants

time   intercept

The rate of a reaction can be measured by dividing the amount
of .... used up or the amount of ....
formed by the .... To find the rate at a particular
time from a graph with a curved line of best fit, you have to
find the .... of the .... at that time.

---

1   To find the rate of a reaction you need to take measurements. Which of the following
could you take measurements of to work out the rate of a reaction?

Grade 4-6

Tick **two** boxes.

Mass ☐   Volume of solution ☐

Volume of gas ☐   Frequency ☐

[Total 2 marks]

---

2   A student wants to investigate how the rate of a particular reaction is affected
by temperature. The reaction produces a precipitate, so she plans to time how
long it takes for the solution to go cloudy at each temperature.

Grade 4-6

2.1   What is the dependent variable in this experiment?

....

[1]

2.2   What is the independent variable in this experiment?

....

[1]

2.3   Suggest **one** variable that would have to be controlled in this experiment to make it a fair test.

....

[1]

2.4   The reaction also produces a gas. State whether it would be more accurate to measure the rate
of the reaction by timing how long it takes for the solution to go cloudy, or by timing how long it
takes a volume of gas to be produced. Explain your answer.

....

....

[2]

[Total 5 marks]

Topic 6 — The Rate and Extent of Chemical Change

3 The rate of a reaction was investigated by measuring the volume of gas produced at regular intervals. The results are shown in **Table 1**.

Table 1

Time (s)	0	60	120	180	240	300	360
Volume of gas (cm^3)	0.0	10.5	15.5	17.4	18.0	18.0	18.0

3.1 Name a piece of equipment that could be used to measure the volume of gas produced.

....

[1]

3.2 Plot the data in **Table 1** on the axes below. Draw a line of best fit onto the graph.

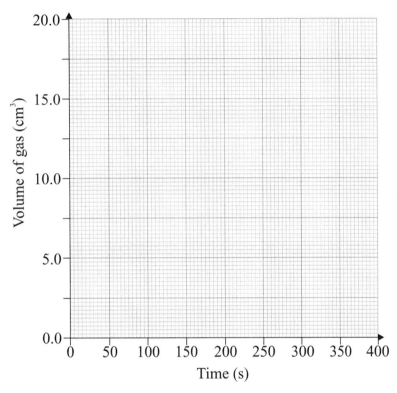

[3]

3.3 At what time does the reaction stop, according to your graph?

....

[1]

3.4 Calculate the mean rate of the reaction.

Mean rate of reaction = ...cm^3/s

[2]

3.5 Suggest how you could check whether the results are repeatable.

....

....

[2]

[Total 8 marks]

Topic 6 — The Rate and Extent of Chemical Change

4 The rate of a reaction between two solutions was investigated by monitoring the amount of one of the reactants, A, at regular intervals. A graph of the results is shown in **Figure 1**.

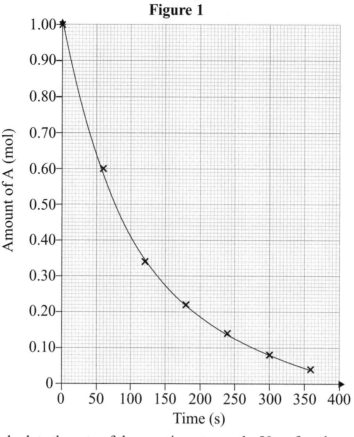

Figure 1

4.1 Use **Figure 1** to calculate the rate of the reaction at exactly 50 s after the start of the reaction. Give this rate to two significant figures. Include units in your answer.

rate = ... units = ...

[4]

4.2 In the same way calculate the rate of the reaction at exactly 200 s after the start of the reaction.

rate = ... units = ...

[4]

4.3 Describe and explain, with reference to collision theory, how the rate changes during the reaction.

....

....

[2]

[Total 10 marks]

Topic 6 — The Rate and Extent of Chemical Change

Reversible Reactions

1 This question is about the chemical reaction that is used to make ammonia from nitrogen and hydrogen. The equation for this reaction is: $N_2 + 3H_2 \rightleftharpoons 2NH_3$ *(Grade 4-6)*

1.1 What does the symbol '$\rightleftharpoons$' mean in this reaction?

....

[1]

1.2 When this reaction is carried out in a sealed container, it reaches equilibrium.
Which of the following statements about equilibrium is true?
Tick **one** box.

At equilibrium, all the reactants have reacted to form products. ☐

At equilibrium, the amount of products equal the amount of reactants. ☐

At equilibrium, the rate of the forward reaction
is equal to the rate of the backwards reaction. ☐

At equilibrium, both the forwards and the backwards reactions stop. ☐

[1]

[Total 2 marks]

2 An aqueous solution of blue copper(II) ions can react with chloride ions to form a yellow copper compound. The ionic equation for this reaction is: $Cu^{2+} + 4Cl^- \rightleftharpoons [CuCl_4]^{2-}$ *(Grade 7-9)*

2.1 The forward reaction is endothermic. What two things can you say about the enthalpy change for the reverse reaction?

....

....

[2]

2.2 A solution containing copper(II) ions is mixed with a solution containing chloride ions in a flask. The solution quickly turns green. When observed for a few minutes no further change in colour can be seen. Explain these observations.

....

....

....

[2]

2.3 Suggest **two** changes that could be made to the mixture that would change its colour.

....

....

[2]

[Total 6 marks]

Topic 6 — The Rate and Extent of Chemical Change

72

Le Chatelier's Principle

Predict whether the changes below would result in there being more reactants or more products at equilibrium for the reaction: $N_{2(g)} + 3H_{2(g)} \rightleftharpoons 2NH_{3(g)}$. The forward reaction is exothermic.

Increasing the temperature: ....

Decreasing the pressure: ....

Increasing the concentration of N_2: ....

1 The position of equilibrium of a reaction is dependent on the conditions that the reaction is carried out under.

Grade 4-6

1.1 What does Le Chatelier's Principle say about the effect of changing the conditions of a reversible reaction at equilibrium?

....
[1]

1.2 State **two** conditions you could change in order to alter the position of equilibrium of a reaction that happens in solution.

....

....
[2]

[Total 3 marks]

2 A mixture of iodine monochloride (ICl) and chlorine is sealed in a gas syringe. The gases react in a reversible reaction to form iodine trichloride (ICl_3) and eventually reach an equilibrium. The equation for the reaction is: $ICl_{(l)} + Cl_{2(g)} \rightleftharpoons ICl_{3(s)}$.

Grade 6-7

2.1 Given that the forward reaction is exothermic, explain how the relative quantities of ICl and ICl_3 would change if the mixture was heated, and all other conditions remained the same.

....

....

....
[2]

2.2 Explain how the relative quantities of ICl and ICl_3 would change if the plunger were pushed into the syringe, and the temperature remained constant.

....

....

....
[3]

[Total 5 marks]

Topic 6 — The Rate and Extent of Chemical Change

3 Dinitrogen tetroxide (N_2O_4) is a colourless gas. It decomposes in a reversible reaction to form the brown gas, nitrogen dioxide (NO_2). The reaction equation is: $N_2O_{4(g)} \rightleftharpoons 2NO_{2(g)}$.

Grade 7-9

3.1 When a sample of N_2O_4 is left to decompose in a sealed tube, a pale brown colour can be seen. If this mixture is heated, the colour becomes a darker brown. Explain this observation and predict whether the forward reaction is exothermic or endothermic.

....

....

....

[3]

3.2 Explain how you would expect the colour of the equilibrium mixture to change if the pressure of the mixture is decreased, and all other conditions are kept the same.

....

....

....

[3]

[Total 6 marks]

4 Yellow iron(III) ions and colourless thiocyanate ions react reversibly in solution to form dark red iron thiocyanate: $Fe^{3+}_{(aq)} + SCN^-_{(aq)} \rightleftharpoons FeSCN^{2+}_{(aq)}$

Grade 7-9

The following observations are made about this reaction:
1. When a yellow solution containing Fe^{3+} ions and a colourless solution containing SCN^- ions are mixed, a pale red colour forms which initially grows darker but then stays constant.
2. When more Fe^{3+} ions are added to the solution it initially becomes more orangey in colour but then grows darker red than before the Fe^{3+} was added and remains like this.
3. If $FeSCN^{2+}$ ions are added to the solution it initially becomes darker in colour but then becomes more orangey.

Explain what is happening in each of these observations.

Observation 1: ....

....

Observation 2: ....

....

Observation 3: ....

....

[Total 6 marks]

Exam Practice Tip

Working out what happens to the position of an equilibrium when you change the conditions can be a bit of a brain twister. Just remember that for any change that's made, the reaction will try to do the opposite. So if you increase the temperature, more of the endothermic reaction will happen, if you increase the pressure, the equilibrium will move to the side where there are fewer moles of gas, and if you increase the concentration of a reactant, you'll get more products.

Topic 6 — The Rate and Extent of Chemical Change

Hydrocarbons

Warm-Up

Place each of the compounds on the left, below in the correct column of the table depending on whether or not they are hydrocarbons.

propane   ethene

butanoic acid

C_2H_6 CH_3CH_2Cl

C_2H_4 hydrochloric acid

Hydrocarbon	Not a hydrocarbon

1   Alkanes are a family of hydrocarbons. (Grade 4-6)

1.1 What is a hydrocarbon?

....
[1]

1.2 Which of the following shows the first four alkanes in order of decreasing carbon chain length? Tick **one** box.

Propane, ethane, butane, methane ☐

Methane, ethane, propane, butane ☐

Ethane, methane, butane, propane ☐

Butane, propane, ethane, methane ☐

[1]

1.3 Write the general formula of alkanes in terms of *n*, where *n* is the number of carbon atoms.

....
[1]

1.4 Complete the word equation for the complete combustion of a hydrocarbon.

hydrocarbon + oxygen → .... + ...
[1]

1.5 During a combustion reaction, are the atoms in the hydrocarbon oxidised or reduced?

....
[1]

[Total 5 marks]

2 The molecular formulas for five hydrocarbons, **A** to **E**, are shown below. Grade 4-6

  A C_4H_8 **B** C_4H_{10} **C** C_5H_{10} **D** C_5H_{12} **E** C_6H_{14}

2.1 Which hydrocarbon is butane?

....
[1]

2.2 Which of the hydrocarbons are alkanes? Explain your answer.

....

....
[2]

2.3 Which of the hydrocarbons is likely to have the highest boiling point? Explain your answer.

....

....
[2]

[Total 5 marks]

3 Petrol and diesel are both fuels containing mixtures of hydrocarbons. The average chain length of the hydrocarbons in petrol and diesel are different, which causes diesel to have a higher boiling point than petrol. Grade 6-7

3.1 Compare the viscosity of petrol and diesel.
Explain your answer with reference to the information above.

....

....
[2]

3.2 Predict whether petrol or diesel will be more flammable.
Explain your answer with reference to the information above.

....

....
[2]

3.3 Diesel contains alkanes that have 20 carbon atoms.
Give the molecular formula of an alkane with 20 carbon atoms.

....
[1]

3.4 Petrol contains alkanes with 8 carbon atoms.
Finish and balance the equation for the complete combustion of this hydrocarbon

.... C_8H_{18} + ... $O_2 \rightarrow$ .... + ...
[2]

[Total 7 marks]

Topic 7 — Organic Chemistry

Fractional Distillation

1 Crude oil is a finite resource. **(Grade 4-6)**

1.1 What is crude oil formed from?

....

[1]

1.2 What does 'finite resource' mean?

....

....

[1]

1.3 What type of substance does crude oil mainly consist of?
Tick **one** box.

Alkenes ☐

Alkanes ☐

Alcohols ☐

Water ☐

[1]

[Total 3 marks]

2 Fractional distillation is used to separate the mixture of molecules in crude oil into fractions such as petrol and diesel oil. **(Grade 6-7)**

2.1 What property of the molecules in crude oil is used to separate them into different fractions?

....

[1]

2.2 Explain how a fractionating column separates the molecules in crude oil into different fractions.

....

....

....

....

[3]

2.3 Fractions boil over a range of temperatures much narrower than the original crude oil.
What does this suggest about the structures of the hydrocarbons in a fraction?

....

[1]

[Total 5 marks]

Topic 7 — Organic Chemistry

Uses and Cracking of Crude Oil

1 Crude oil is processed to be used for a variety of purposes. (Grade 4-6)

1.1 Suggest **two** types of useful material produced from crude oil fractions.

....

[2]

1.2 Long-chain hydrocarbons can be processed to produce short-chain hydrocarbons.
What is the name of this process?
Tick **one** box.

Fractional distillation ☐

Thermal reduction ☐

Thermal oxidation ☐

Cracking ☐

[1]

1.3 Why do we break up long chain hydrocarbons into shorter chain hydrocarbons?

....

[1]

[Total 4 marks]

2 Some hydrocarbons from crude oil undergo processing by the petrochemical industry. For instance, decane, $C_{10}H_{22}$, can undergo cracking as shown in the following equation: (Grade 6-7)
$$C_{10}H_{22} \rightarrow C_8H_{18} + C_2H_4$$

2.1 What type of reaction is cracking?

....

[1]

2.2 Describe how cracking is carried out.

....

....

[2]

2.3 Cracking can form a variety of products.
Write an alternative balanced equation for the cracking of decane.

....

[1]

[Total 4 marks]

Topic 7 — Organic Chemistry

3   When large alkanes are cracked, smaller alkanes and alkenes are produced. An example of cracking is given in **Figure 1**.

Figure 1

H–C–C–C–C–C–C–C–H ⟶ [C structure] + **D**

C

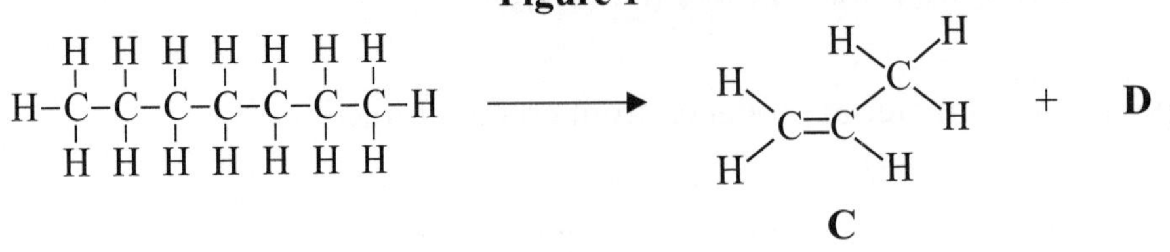

3.1 Write the chemical formula of the reactant.

....

[1]

3.2 Draw the displayed formula of **D**.

[2]

3.3 Suggest a use for the alkene, **C**.

....

[1]

[Total 4 marks]

4*   Explain why and how some fractions of crude oil are processed by cracking, giving chemical equations where relevant. You should state some uses of the products of this process.

....

....

....

....

....

....

....

....

....

....

[Total 6 marks]

Exam Practice Tip

When balancing cracking equations, make sure that there's at least one alkane and alkene in the products. Also, just like any chemical equation, remember to check that the number of each atom on both sides of the equations is the same.

Alkenes

1 Alkenes form a homologous series of reactive organic compounds.

1.1 Which of the following is the most accurate description of an alkene?
Tick **one** box.

A hydrocarbon with a single carbon-carbon bond ☐

A hydrocarbon with a double carbon-carbon bond ☐

A hydrocarbon with a triple carbon-carbon bond ☐

A hydrocarbon with a double carbon-hydrogen bond ☐

[1]

1.2 What is the general formula for alkenes?

....
[1]

1.3 Draw the displayed formula of propene, C_3H_6.

[1]
[Total 3 marks]

2 Five organic compounds, **A** to **E** are shown in **Figure 1**.

Figure 1

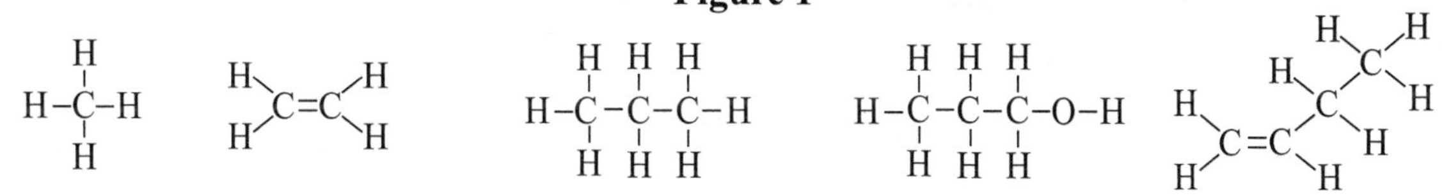

| **A** | **B** | **C** | **D** | **E** |

2.1 Which of the molecules are unsaturated?

....
[1]

2.2 Give the molecular formula of a molecule in the same homologous
series as **E** that contains 6 carbon atoms.

....
[1]

2.3 In low amounts of oxygen, **B** will burn with a smoky flame. State what type of combustion this is
and give a balanced symbol equation for the combustion of **B** in low amounts of oxygen.

....

....
[3]
[Total 5 marks]

Topic 7 — Organic Chemistry

Reactions of Alkenes

Fill in the gaps for the following paragraph using the words below on the left.

alcohols   addition

bromine water

unsaturated

ethene

Alkenes generally react via .... reactions to form a variety of compounds. Alkenes can react with steam to form .... For example, ... can be mixed with steam and passed over a catalyst to form ethanol.

1 Ethene is a hydrocarbon that takes part in addition reactions.

1.1 What is the functional group in ethene?

....

[1]

1.2 Why do other compounds in the same homologous series as ethene react in similar ways?

....

[1]

1.3 Describe what happens to the functional group in ethene during an addition reaction.

....

....

[2]

[Total 4 marks]

2 Alkenes can react with hydrogen gas to form alkanes.
The reaction between propene and hydrogen is shown in **Figure 1**.

Figure 1

$$\underset{H}{\overset{H}{>}}C=C\underset{H}{\overset{H}{<}}\overset{H}{\underset{H}{C}}\overset{H}{<}_{H} \quad + \quad H_2 \quad \longrightarrow \quad H-\overset{H}{\underset{H}{C}}-\overset{H}{\underset{H}{C}}-\overset{H}{\underset{H}{C}}-H$$

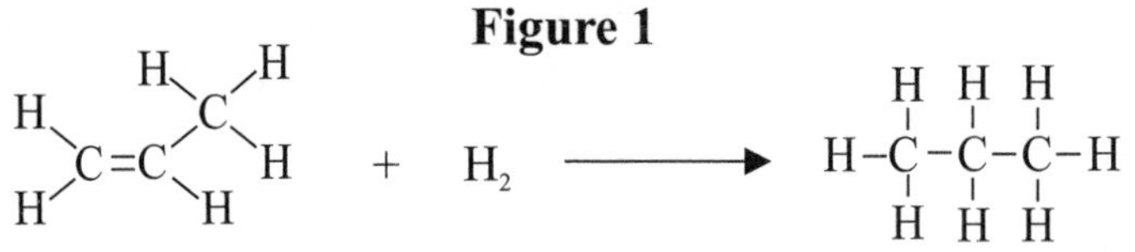

2.1 What is the molecular formula of propene?

....

[1]

2.2 What substance is needed to increase the rate of reaction?

....

[1]

2.3 Describe a simple chemical test that would allow you to distinguish between the organic reactant and product. Describe what you would observe in each case.

....

....

....

[3]

[Total 5 marks]

3 Alkenes can react in an addition reaction with halogens. An example of the reactants that can take part in this type of reaction is shown in **Figure 2**.

Figure 2

H H H H
H-C-C=C-C-H... + Br₂ ⟶ X

(displayed formula of but-2-ene shown) + Br₂ ⟶ X

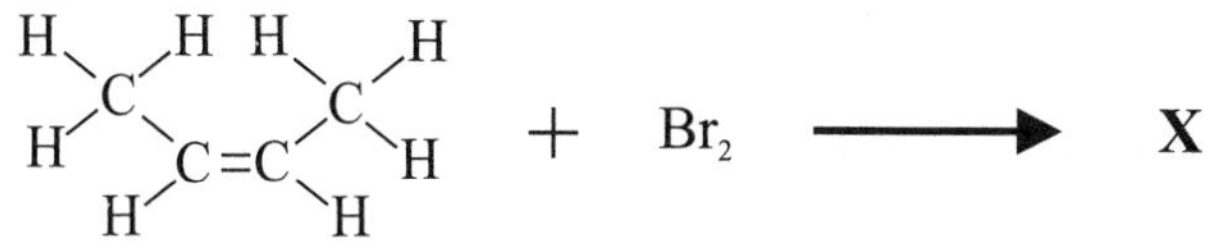

3.1 Draw the displayed formula of product **X**.

[1]

3.2 Is the product of this reaction saturated or unsaturated? Explain your answer.

....

....

[2]

[Total 3 marks]

4 Propene is a feedstock for the production of many useful organic compounds.

4.1 Interhalogen compounds such as ICl contain 2 different halogen atoms. They react with alkenes in a similar way to halogens. Draw the displayed formula of the product of a reaction between ICl and propene.

[1]

4.2 Propene also reacts with steam in the presence of a catalyst. There are two possible products of this reaction. Draw the displayed formulas of the two different products.

[2]

[Total 3 marks]

Exam Practice Tip

Don't get in a tizz if you're asked about the reaction of an alkene with a reactant you've never seen before. All you need to remember is that the double bond in the alkene is likely to react in the same way as it does in all the reactions you've met. So keep a cool head, think about the general rules of alkene addition reactions and you'll be sorted.

Topic 7 — Organic Chemistry

Addition Polymers

1   Poly(ethene) is a polymer used in packaging applications.   (Grade 4-6)

1.1   Name the monomer used to form poly(ethene).

....

[1]

1.2   What type of polymerisation reaction occurs to form poly(ethene) from its monomer?

....

[1]

1.3   What functional group is involved in the formation of poly(ethene) from its monomer?

....

[1]

[Total 3 marks]

2   A short length of the polymer chain of poly(chloroethene) is shown in **Figure 1**.   (Grade 6-7)

Figure 1

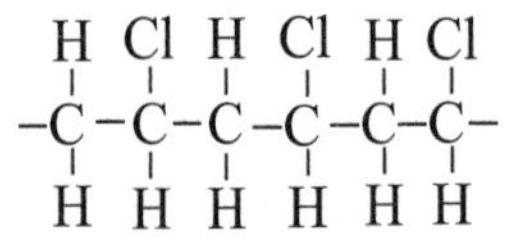

```
   H  Cl H  Cl H  Cl
   |  |  |  |  |  |
 –C –C –C –C –C –C–
   |  |  |  |  |  |
   H  H  H  H  H  H
```

2.1   Draw the repeating unit for the polymer chain in **Figure 1**.

[1]

2.2   Draw the displayed formula of the monomer.

[1]

2.3   Name the monomer used to form poly(chloroethene).

....

[1]

[Total 3 marks]

Exam Practice Tip

Don't panic if you're asked to find the repeating unit from a polymer chain. The carbon backbone in the repeating unit of an addition polymer only ever has two carbon atoms. So start by drawing these two carbon atoms and then look at the polymer chain to work out what groups surround them. Add brackets around the end bonds, a little 'n', and voilà.

Topic 7 — Organic Chemistry

Alcohols

1   This question is about alcohols.   (Grade 4-6)

   1.1   State the functional group present in alcohols.

     ....

[1]

   1.2   State **two** uses of alcohols.

     ....

[2]

   1.3   A few drops of methanol were added to test tube containing Universal indicator solution. What would you observe? Tick **one** box.

     Two layers since methanol does not dissolve in water. The indicator remains green. ☐

     Methanol dissolves. The indicator remains green. ☐

     Methanol dissolves. The indicator turns orange. ☐

     Methanol dissolves. The indicator turns blue. ☐

[1]

   1.4   What are the products when an alcohol is completely combusted? Tick **one** box.

     CO and H_2O ☐

     CO_2 and H_2O ☐

     CO and H_2 ☐

     CO_2 and H_2 ☐

[1]

[Total 5 marks]

2   Ethanol is used as a starting material for other organic products. It can be made by fermenting sugar.   (Grade 6-7)

   2.1   Ethanol can be oxidised. Give the name and type of organic compound it is oxidised to.

     ....

[2]

   2.2   State the reactants and conditions needed to ferment sugar solutions to form ethanol.

     ....

     ....

     ....

[4]

[Total 6 marks]

Topic 7 — Organic Chemistry

3   Methanol and butanol are both alcohols.

3.1 Draw the displayed formula of methanol.

[1]

3.2 Give one similarity and one difference between the structures of methanol and butanol.

....

....

[2]

3.3 Why do methanol and butanol both react in similar ways?

....

[1]

3.4 Both methanol and butanol can react with sodium. One of the products of the reaction is gaseous. Name the gaseous product in the reaction of methanol and butanol with sodium.

....

[1]

[Total 5 marks]

4   Ethane-1,2-diol is used as coolant, as antifreeze in car radiators and in the production of other organic compounds. The displayed formula for ethane-1,2-diol is shown in **Figure 1**.

Figure 1

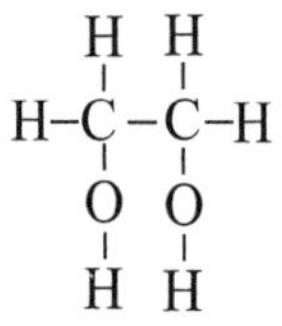

$$\begin{array}{ccc} H & H \\ | & | \\ H-C- & C-H \\ | & | \\ O & O \\ | & | \\ H & H \end{array}$$

4.1 Write the formula of ethane-1,2-diol.

....

[1]

4.2 How could you produce hydrogen from ethane-1,2-diol?

....

[1]

4.3 Write a balanced equation for the complete combustion of ethane-1,2-diol.

....

[2]

4.4 Predict the pH of an aqueous solution of ethane-1,2-diol.

....

[1]

[Total 5 marks]

Carboxylic Acids

1   Methanoic acid is the simplest carboxylic acid possible.   (Grade 4-6)

1.1   What is the functional group in methanoic acid?
Tick **one** box.

$-CH_3$ ☐

$C=C$ ☐

$-OH$ ☐

$-COOH$ ☐

[1]

1.2   Describe what would be observed if a few drops of methanoic acid were added to Universal indicator solution in a test tube.

....

....

[2]

1.3   Methanoic acid reacts with calcium carbonate. What gas is evolved?

....

[1]

[Total 4 marks]

2   A carboxylic acid is shown in **Figure 1**.
It reacts with ethanol to form another organic compound.   (Grade 6-7)

Figure 1

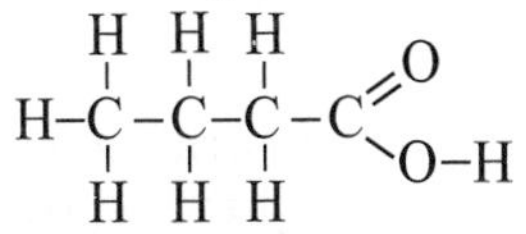

$$H-\overset{\overset{\displaystyle H}{|}}{\underset{\underset{\displaystyle H}{|}}{C}}-\overset{\overset{\displaystyle H}{|}}{\underset{\underset{\displaystyle H}{|}}{C}}-\overset{\overset{\displaystyle H}{|}}{\underset{\underset{\displaystyle H}{|}}{C}}-C\overset{\nearrow O}{\searrow_{O-H}}$$

2.1   Name the carboxylic acid and write its formula.

....

[2]

2.2   Explain why the carboxylic acid shown is a weak acid.

....

[1]

2.3   What additional substance would be needed to increase the reaction rate in the reaction between the compound in **Figure 1** and ethanol?

....

[1]

2.4   What type of compound is made by the reaction between the compound in **Figure 1** and ethanol?

....

[1]

[Total 5 marks]

3   Ethanoic acid is a weak acid found in vinegar.   Grade 6-7

3.1   Estimate the pH of a solution of ethanoic acid.

....

[1]

3.2   A solid compound, **R**, which contained magnesium ions, was added to vinegar in a conical flask. Carbon dioxide was evolved. Identify **R**.

....

[1]

3.3   Ethanoic acid was reacted with organic compound **T** in the presence of an acid catalyst, forming ethyl ethanoate. Identify **T**.

....

[1]

[Total 3 marks]

4   Propanoic acid is a carboxylic acid with the molecular formula $C_3H_6O_2$.   Grade 7-9

4.1   Work out the formula of **F** in the following equation:

$$2CH_3CH_2COOH + F \rightarrow 2CH_3CH_2COO^- Na^+ + H_2O + CO_2$$

....

[1]

4.2   Propanoic acid reacts with alcohols to form esters.
Give the formula of **W** in the following equation:

$$CH_3CH_2COOH + CH_3CH_2OH \rightarrow CH_3CH_2COOCH_2CH_3 + W$$

....

[1]

4.3   The percentage ionisation of two acids was determined. Both acids were the same concentration. Acid **D** was found to be 1.1% ionised and acid **E** was 98% ionised. Determine which acid is propanoic acid and justify your answer.

....

....

....

[3]

4.4   Propanoic acid can burn in oxygen. Give the products of the reaction when propanoic acid undergoes complete combustion.

....

[2]

[Total 7 marks]

Topic 7 — Organic Chemistry

Condensation Polymers

1   Condensation polymerisation is used to produce many different polymers.   Grade **4-6**

1.1 Which of the following is true about condensation polymerisation?
Tick **one** box.

The monomers have carbon-carbon double bonds.   ☐

The repeat unit of the polymer has the same atoms as the combined monomers.   ☐

A small molecule is lost when condensation polymers are formed.   ☐

A small molecule is gained when condensation polymers are formed.   ☐
[1]

1.2 How many functional groups does each monomer need to
have to undergo condensation polymerisation?

....
[1]

1.3 Name **two** functional groups that can react together in condensation polymerisation reactions.

....
[2]

[Total 4 marks]

2   Ethane-1,2-diol and hexanedioic acid polymerise to produce a polyester, **D**.   Grade **6-7**
The reactants are shown in simplified form in **Figure 1**.

Figure 1

n HO—▨—OH + n HOOC—☐—COOH ⟶   **D**   +   2n **E**

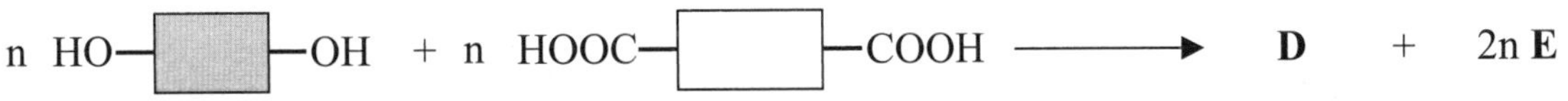

2.1 What is the formula of **E**?

....
[1]

2.2 Draw the repeat unit of polyester **D** in the simplified form.

[2]

2.3 Would it be possible to form a polymer from just one of the monomer reactants in **Figure 1**?
Explain your answer.

....

....
[2]

[Total 5 marks]

😕 ☐   🙂 ☐   😃 ☐   Topic 7 — Organic Chemistry

Naturally Occurring Polymers

1 Glycine is the simplest of the 20 naturally occurring amino acids. It is shown in **Figure 1**.

Figure 1

$$H_2N-\underset{\underset{H}{|}}{\overset{\overset{H}{|}}{C}}-COOH$$

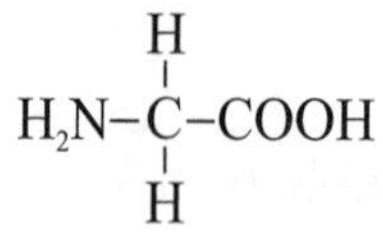

1.1 How many functional groups are there in an amino acid?

....

[1]

1.2 Name **one** of the functional groups present in glycine.

....

[1]

1.3 What type of reaction results in the formation of polypeptides from amino acids?

....

[1]

1.4 State the name of the small molecule lost during the formation of polypetides.

....

[1]

[Total 4 marks]

2 Starch and DNA are both naturally occurring polymers.

2.1 What type of molecule are the monomers in starch?

....

[1]

2.2 Name **one** naturally occurring polymer that is made up of the same type of monomers as starch.

....

[1]

2.3 What is the role of DNA?

....

....

[2]

2.4 Describe the general structure of DNA.

....

....

....

[3]

[Total 7 marks]

Topic 7 — Organic Chemistry

Purity and Formulations

1 Copper can be made extremely pure. The melting point of two samples of copper were measured. Sample **A** had a melting point of 1085 °C and sample **B** melted over the range 900 – 940 °C.

1.1 How is a pure substance defined in chemistry? Tick **one** box.

A single element not mixed with any other substance. ☐

A single compound not mixed with any other substance. ☐

A single element or compound not mixed with any other substance. ☐

An element that has not been reacted with anything. ☐

[1]

1.2 Suggest which of the samples was the most pure? Explain your answer.

....

....

[2]

1.3 The boiling point of copper is 2,562 °C. Which of the samples is likely to have a boiling point closer to that of pure copper?

....

[1]

[Total 4 marks]

2 A paint was composed of 20% pigment, 35% binder, 25% solvent, and 20% additives. **Grade 6-7**

2.1 Explain why the paint is a formulation.

....

....

....

[3]

2.2 How would a manufacturer of the paint ensure that each batch had exactly the same properties?

....

[1]

2.3 Other than paint, name **one** other example of a formulation.

....

[1]

[Total 5 marks]

Exam Practice Tip

A formulation is a mixture but a mixture isn't always a formulation. For the exam, you may need to identify formulations based on information about their ingredients and how they've been designed to make the product fit for purpose.

 ☐   ☐   ☐

Paper Chromatography

Warm-Up

Complete the diagram by correctly labelling the different parts of the chromatography apparatus using the labels on the left, below.

Watch glass

Sample

Filter paper

Baseline

Solvent

Spots of chemicals

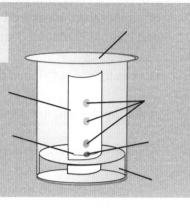

1   Paper chromatograms were produced for three dyes, **D**, **E** and **F**, using a variety of solvents. The chromatogram produced using ethanol as a solvent is shown in **Figure 1**.

Figure 1

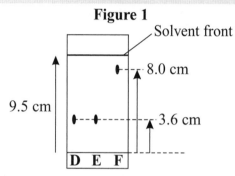

Solvent front

8.0 cm

9.5 cm

3.6 cm

D   E   F

1.1   Calculate the R_f values for **E** and **F** in ethanol, using the chromatogram shown in **Figure 1**.

R_f of **E** = ...

R_f of **F** = ...
[4]

1.2   Why do the substances travel different distances?

....

....
[1]

1.3   In all solvents, each dye only has one spot.
      What does this imply about the composition of the dyes?

....
[1]

1.4   State which of the dyes could be the same.

....
[1]

[Total 7 marks]

2 A paper chromatography experiment was used to identify the compounds in a mixture, **W**, as shown in **Figure 2**. Three known compounds, **A**, **B** and **C** were spotted alongside **W**. Water was used as the solvent.

Grade 6-7

2.1 The experiment was conducted in a beaker with a watch glass functioning as a lid. Why is it necessary to have a lid while conducting a paper chromatography experiment?

....

[1]

2.2 The R_f values of **A** and **B** were found to be 0.9 and 0.1 respectively. With reference to the mobile phase and stationary phase, suggest why **A** has a much larger R_f value than **B**.

....

[1]

2.3 A diagram of the chromatogram from the experiment can be seen in **Figure 2**. State which of the known compounds could be found in **W**.

Figure 2

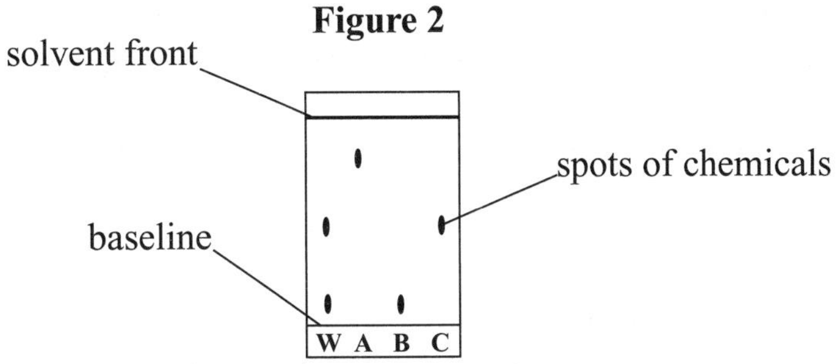

solvent front

spots of chemicals

baseline

W A B C

....

[1]

2.4 A student suggested that if the solvent was changed, the R_f value of compounds **A**, **B** and **C** would stay the same. Explain whether the student was correct.

....

....

....

[2]

2.5 The experiment was repeated again using a different solvent. The mixture **W** had three spots on the resultant chromatogram. What does this suggest about the mixture **W**?

....

[1]

2.6 Suggest why only two spots were shown on the chromatogram shown in **Figure 2**.

....

....

[1]

[Total 7 marks]

Exam Practice Tip

There are lots of different type of chromatography, but paper chromatography is the only one you need to know how to carry out. It's not just about knowing how to set up the apparatus — you also need to know how to analyse the results.

Topic 8 — Chemical Analysis

Tests for Gases and Anions

1 This question is about common tests for gases. *Grade 4-6*

1.1 What is the test for hydrogen? Tick **one** box.

☐ A burning splint which results in a popping noise.

☐ A glowing splint which results in a popping noise.

☐ A burning splint which is extinguished.

[1]

1.2 The gas produced in a reaction was collected in a test tube. A glowing splint relit when inserted into the tube. Identify the gas given off.

....

[1]

1.3 Outline a test for carbon dioxide gas. Include any observations you would expect from a positive result.

....

....

[2]

1.4 How could you test for chlorine gas? Include any observations you would expect from a positive result.

....

....

[2]

[Total 6 marks]

2 Certain positive ions can be identified using flame tests. *Grade 4-6*

2.1 A flame test was conducted on a sample containing an unknown cation. The sample burned with a lilac flame. Identify the unknown cation. Tick **one** box.

☐ Lithium, Li^+ ☐ Potassium, K^+ ☐ Copper, Cu^{2+} ☐ Calcium, Ca^{2+}

[1]

2.2 Describe how to conduct a flame test.

....

....

....

....

[4]

2.3 A solution containing sodium ions was tested using a flame test. What colour flame would you expect to see?

....

[1]

[Total 6 marks]

3 A student carried out tests to identify the compound present in a solution. **Grade 6-7**

3.1 The student added a couple of drops of hydrochloric acid
followed by a couple of drops of barium chloride to the solution.
A white precipitate was formed. Suggest what anion was present in the solution.

....
[1]

3.2 Write a balanced ionic equation for the formation of the white precipitate. Include state symbols.

....
[2]

3.3 The student added a few drops of sodium hydroxide solution to the solution containing the
compound. A blue precipitate was formed. Suggest what cation was present in the solution.

....
[1]

3.4 Use your answers to 3.1 and 3.3 to identify the compound in the solution.

....
[1]

[Total 5 marks]

4 Some simple chemical tests were carried out on three unknown substances
in solution, **P**, **Q** and **R**. The results are shown in the **Table 1**. **Grade 7-9**

Table 1

Substance	Flame test colour	Addition of sodium hydroxide, $NaOH_{(aq)}$	Addition of silver nitrate $(AgNO_3)_{(aq)}$ in the presence of nitric acid
P	Orange-red	**D**	yellow precipitate
Q	**E**	Blue precipitate	no reaction
R	Crimson	No precipitate	cream precipitate

4.1 Identify substances **P** and **R**.

Substance **P**: ....

Substance **R**: ....
[4]

4.2 Why is it not possible to identify substance **Q**?

....
[1]

4.3 State the observation for **D** and **E**.

D: ....

E: ....
[2]

4.4 **P** and **Q** were mixed in equal quantities.
Why would it be difficult to use a flame test to identify the ions in this mixture?

....
[1]

[Total 8 marks]

Topic 8 — Chemical Analysis

Flame Emission Spectroscopy

1 Flame emission spectroscopy is an example of an instrumental method that can be used to analyse elements and compounds.

Grade 6-7

1.1 State **two** advantages of using instrumental methods compared to chemical tests.

....

....

[2]

1.2 What two pieces of information can flame emission spectroscopy tell you about metal ions in solution?

....

....

[2]

1.3 What type of spectra does flame emission spectroscopy produce?

....

[1]

[Total 5 marks]

2 Flame emission spectroscopy can be used to detect various positive ions within a mixture. **Figure 1** shows the spectra for Metal **A**, Metal **B**, Metal **C** and a mixture, **M**.

Grade 7-9

Figure 1

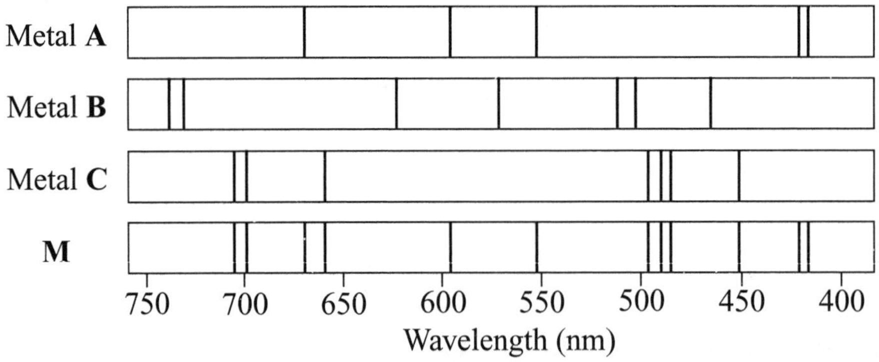

Metal **A**

Metal **B**

Metal **C**

M

750 700 650 600 550 500 450 400

Wavelength (nm)

2.1 Describe how flame emission spectroscopy works.

....

....

....

[2]

2.2 Which metal ion(s) are in the mixture, **M**?

....

[1]

[Total 3 marks]

Topic 8 — Chemical Analysis

The Evolution of the Atmosphere

Warm-Up

Circle whether each of the following sentences is **true** or **false**.

1) Earth's early atmosphere contained mainly nitrogen.   True   or   False

2) Earth's early atmosphere formed from gases released by volcanoes.   True   or   False

3) Earth's early atmosphere contained very little oxygen.   True   or   False

4) The oceans formed when oxygen reacted with hydrogen to form water.   True   or   False

1   The composition of gases in the atmosphere has varied during Earth's history.   **Grade 4-6**

1.1   What are the approximate proportions of oxygen and nitrogen in the atmosphere today?
       Tick **one** box.

       One-fifth oxygen and four-fifths nitrogen.   ☐

       Two-fifths oxygen and three-fifths nitrogen.   ☐

       Three-fifths oxygen and two-fifths nitrogen.   ☐

       Four-fifths oxygen and one-fifth nitrogen.   ☐

                                                                                    [1]

1.2   Other than oxygen and nitrogen, name **two** other gases in the atmosphere today.

       ....
                                                                                    [2]

1.3   How was the oxygen in the atmosphere produced?

       ....
                                                                                    [1]

1.4   How was nitrogen in the atmosphere produced?

       ....
                                                                                    [1]

1.5   For approximately how long has the atmosphere had a composition similar to what it is today?
       Tick **one** box.

       5 million years ☐   30 million years ☐   200 million years ☐   1 billion years ☐

                                                                                    [1]
                                                                          [Total 6 marks]

2 During the Earth's first billion years, the percentage of carbon dioxide in the atmosphere was probably much higher than it is today. The formation of sedimentary rocks and fossil fuels helped to decrease the percentage of carbon dioxide.

Grade 6-7

2.1 Suggest **one** reason, other than sedimentary rock and fossil fuel formation, why the percentage of carbon dioxide in the atmosphere is thought to have decreased.

....

[1]

2.2 Outline how sedimentary rocks are formed.

....

....

[1]

2.3 State what the sedimentary rocks coal and limestone are formed from.

Coal: ....

Limestone: ....

[2]

[Total 4 marks]

3 There are several theories about how Earth's atmosphere evolved to its current composition. One theory suggests that the proportion of oxygen in Earth's early atmosphere was similar to the proportion in Mars' atmosphere today, which is 0.13%. The theory then suggests that, about 2.7 billion years ago, the proportion of oxygen in Earth's atmosphere started increasing.

Grade 7-9

3.1 Suggest why it's difficult to come up with a conclusive theory about how the atmosphere evolved.

....

[1]

3.2 Write a balanced symbol equation for the reaction that is thought to have caused the proportion of oxygen in Earth's atmosphere to increase.

....

[1]

3.3 Suggest why there is such a small amount of oxygen in the atmosphere of Mars.

....

....

[2]

3.4 Red beds are rocks that contain iron oxide. They form when other iron compounds from older rocks come into contact with oxygen in the atmosphere, and react. The oldest red beds formed about 2 billion years ago.

Suggest how this provides evidence for when the amount of oxygen in Earth's atmosphere started increasing.

....

....

[1]

[Total 5 marks]

Topic 9 — Chemistry of the Atmosphere

Greenhouse Gases and Climate Change

1 Greenhouse gases in the atmosphere help maintain life on Earth. (Grade 4-6)

1.1 Which of the following is **not** a greenhouse gas?
Tick **one** box.

Carbon dioxide ☐   Methane ☐

Nitrogen ☐   Water vapour ☐

[1]

1.2 State how greenhouse gases help to support life on Earth.

....
[1]

1.3 Give **two** examples of types of human activity which are leading to an increase in the concentration of greenhouse gases in the atmosphere.

....

....
[2]

[Total 4 marks]

2 An increase in the amount of greenhouse gases in the atmosphere may cause the climate to change. (Grade 6-7)

2.1 Explain how greenhouse gases help to keep the Earth warm.
Your answer should make reference to the interaction of greenhouse gases with radiation.

....

....

....
[3]

2.2 An increase in global temperatures could cause sea levels to rise.
Give **two** potential consequences of rising sea levels.

....

....
[2]

2.3 Other than rising sea levels, give **one** other potential consequence of climate change.

....
[1]

[Total 6 marks]

Topic 9 — Chemistry of the Atmosphere

3* The global temperature anomaly is the difference between current temperature and an average value. The graph in **Figure 1** shows how the global temperature anomaly and the concentration of CO_2 in the atmosphere have varied over time. Describe the trends in the data and suggest reasons for them.

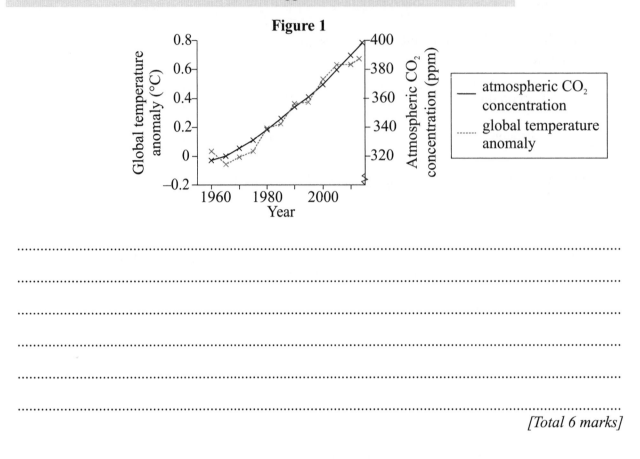

Figure 1

....

....

....

....

....

....

[Total 6 marks]

4 **Table 1** shows some data about atmospheric greenhouse gases. The global warming potential measures the impact of the gas on global warming compared to carbon dioxide.

Table 1

Gas	Formula	Current atmospheric level (%)	Lifetime in atmosphere (years)	Global warming potential
Carbon dioxide	CO_2	0.040	100-300	1
Methane	CH_4	0.00019	12	28
CFC-12	CCl_2F_2	5.3×10^{-7}	100	10 200

4.1 Use the table to suggest why methane is an important greenhouse gas, despite currently being at a lower level than carbon dioxide.

....

[1]

4.2 Use data from the table to explain why CFC-12 could have a significant impact on global warming.

....

....

[2]

[Total 3 marks]

Topic 9 — Chemistry of the Atmosphere

Carbon Footprints

1 Individuals have an annual carbon footprint. One of the factors that contributes to each person's carbon footprint is the amount of electrical energy they use each year.

Grade 4-6

1.1 What is meant by the term 'carbon footprint'?

....

....

[2]

1.2 Suggest **two** ways that the carbon footprint resulting from an individual's electricity usage could be reduced.

....

....

[2]

1.3 Suggest why an individual might not try to reduce their carbon footprint.

....

[1]

[Total 5 marks]

2 In recent years, some businesses have tried to reduce their carbon footprint by capturing the carbon dioxide they produce and storing it, rather than releasing it into the atmosphere.

Grade 6-7

2.1 Suggest **two** disadvantages for companies of capturing and storing carbon dioxide.

....

....

[2]

2.2 Discuss **two** ways in which governments can encourage businesses to reduce their carbon footprints and explain why some countries may be resistant to using these methods.

....

....

....

....

....

[4]

[Total 6 marks]

Exam Practice Tip

Learning the definitions for all the different terms that crop up in GCSE chemistry may be a bit of a bore, but it might be really useful in the exams. Learning all the itty bitty details is worth it if it means you get all the marks available.

Topic 9 — Chemistry of the Atmosphere

Air Pollution

1 A variety of pollutants can be released when fuels burn, with a range of consequences. *(Grade 4-6)*

1.1 Why can the combustion of coal produce sulfur dioxide?

....

[1]

1.2 Name one pollutant that can lead to each of the following consequences:

Acid rain: ....

Global dimming: ....

[2]

1.3 State **two** ways in which acid rain can be damaging.

....

....

[2]

[Total 5 marks]

2 Combustion of fuel in cars is a major contributor to air pollution. *(Grade 6-7)*

2.1 Explain how cars produce nitrogen oxides.

....

....

[2]

2.2 Give **two** problems caused by nitrogen oxides in the environment.

....

....

[2]

2.3 Fuel combustion can produce particulates. What impact do particulates have on human health?

....

[1]

2.4 Combustion of fuels can also produce a gas that prevents blood from carrying oxygen around the body. Inhaling it can cause health problems, and sometimes death.

Name the gas and state why it is difficult to detect it.

....

....

[2]

[Total 7 marks]

Ceramics, Polymers and Composites

1 There are many different types of ceramic with very different properties. For example, brick and glass are both ceramic materials. **(Grade 4-6)**

1.1 What three ingredients are used to make soda-lime glass?
Tick **three** boxes.

☐ Limestone ☐ Sand ☐ Salt ☐ Sodium hydroxide ☐ Sodium carbonate

[3]

1.2 Why are boiling tubes, used for heating liquids, made from borosilicate glass rather than soda-lime glass?

....

[1]

1.3 Briefly describe how bricks are made from clay.

....

....

[2]

[Total 6 marks]

2 This question is about the structures of polymers and composites. **(Grade 6-7)**

2.1 Low density poly(ethene) and high density poly(ethene) are both made from the same starting material, ethene. What causes them to have different properties?

....

....

[2]

2.2 Poly(ethene) melts when it is heated. Is it a thermosetting or thermosoftening polymer?

....

[1]

2.3 Polyester resin is a plastic that does not melt when it is heated. Explain how the structures of poly(ethene) and polyester resin cause them to have different properties when heated.

....

....

....

[2]

2.4 Polymers can be used to make composite materials.
Describe the general structure of a composite.

....

....

[2]

[Total 7 marks]

Properties of Materials

1   This question is about mixtures of different metals known as alloys.

1.1   Draw **one** line between each alloy and the mixture of elements that it contains.

| Bronze | | Copper and zinc |

| Steel | | Copper and tin |

| | Iron and carbon |

| Brass | | Aluminium and copper |

[3]

1.2   State **one** use for the alloy brass.

....

[1]

[Total 4 marks]

2   This question is about different alloys and their uses.
     Table 1 shows some data about the properties of various alloys.

Table 1

Alloy	Carbon Composition (%)	Strength (MPa)	Density (g/cm³)
Stainless steel	0.08	205	8.03
Low carbon steel	0.1	245	7.60
High carbon steel	1.5	355	7.84
Aluminium alloy	0	117	2.71

2.1   Use the data in **Table 1** to state the effect on strength of increasing the carbon content in steel.

....

[1]

2.2   A vice is used to hold an object in place while work is carried out on it.
     The material a vice is made from needs to be strong and heavy to hold objects in place.
     Suggest an alloy from **Table 1** which would be suitable for this purpose.

....

[1]

2.3   Aluminium alloy has the lowest strength value of all the metals shown in the table and yet it is
     used to make many parts of commercial aircraft. Using the data in **Table 1**, explain why this is.

....

[1]

[Total 3 marks]

Topic 10 — Using Resources

Corrosion

1 Iron reacts with water and oxygen to form hydrated iron(III) oxide, known as rust. **Grade 6-7**

1.1 Write the word equation for the formation of rust.

....

[1]

1.2 Iron can be protected by a process known as galvanising. Describe what galvanising is and state the advantage of galvanising over other methods of protection.

....

....

....

[3]

1.3 Suggest **three** methods other than galvanising that could be used to prevent iron from rusting.

....

....

[3]

[Total 7 marks]

2 This question is about corrosion in metals and how to prevent it. **Grade 6-7**

2.1 What does the term 'corrosion' mean?

....

....

[1]

2.2 Why isn't corrosion a problem for things made of aluminium?

....

....

....

[2]

2.3 Magnesium is more reactive than iron. Blocks of magnesium can be attached to the steel in a ship's hull to prevent corrosion. Briefly explain why this technique works.

....

....

[1]

[Total 4 marks]

Finite and Renewable Resources

1 This question is about sustainable use of the Earth's resources. **Table 1** shows the time it takes to form various materials.

Table 1

Material	Time to form (years)
Wood	2-20
Coal	3×10^8
Cotton	0.5

1.1 Using the data in **Table 1**, state **one** finite resource. Explain your answer.

....

....

[2]

1.2 What is meant by the term 'renewable resource'?

....

[1]

[Total 3 marks]

2 Humans have developed items made from both natural and synthetic materials.

2.1 Give **one** example of how agriculture is used to increase
the supply of an otherwise natural resource.

....

....

[1]

2.2 Give **one** example of a synthetic product which has replaced
or is used in addition to a natural resource.

....

[1]

[Total 2 marks]

3 The mining of metals has economic, environmental and social impacts.

Give **one** advantage and **one** disadvantage of metal mining.

....

....

[Total 2 marks]

Reuse and Recycling

1   This question is on sustainable development.   (Grade 6-7)

1.1   What is sustainable development?

....

....

[2]

1.2   How do chemists play a role within sustainable development?
Explain your answer using an example.

....

....

....

[2]

[Total 4 marks]

2   Carrier bags can be made from a number of different materials. Two possible materials   (Grade 6-7)
are jute and poly(ethene). **Table 1** gives some information about these materials.

Table 1

	Poly(ethene)	Jute
Source	Crude oil	Plant fibre
Energy of Production	Moderate energy production	High energy production
Biodegradability	Not biodegradable	Biodegradable
Recyclability	Can be recycled	Possible but not widely done, likely to be reused

2.1   Using the information in **Table 1**, compare the sustainability
of the raw materials needed to make the bags.

....

....

[2]

2.2   Using the information in **Table 1**, compare the sustainability of the production of the bags.

....

....

[2]

2.3   Using the information in **Table 1**, compare the sustainability of the disposal of both bags.

....

....

....

[3]

[Total 7 marks]

3 This question is on recycling. **Grade 6-7**

3.1 Give **two** reasons why recycling can be more sustainable than making new materials.

....

....

[2]

3.2 Name a material that is commonly recycled and briefly describe the process.

....

....

[2]

3.3 State **one** alternative to recycling that also improves sustainability.

....

[1]

[Total 5 marks]

4 This question is about the sustainability of copper. **Grade 7-9**

4.1 Explain how the process of phytomining is used to produce a substance
    which contains copper compounds.

....

....

....

....

[4]

4.2 Describe **two** ways in which copper metal can be extracted from the product of phytomining.

....

....

[2]

4.3 Explain why phytomining cannot ultimately provide a sustainable source of copper metals.
    Suggest **one** thing that can be done to make use of copper more sustainable.

....

....

....

[3]

[Total 9 marks]

Exam Practice Tip

Make sure you understand how reuse and recycling are used to make the materials we use in our daily lives
more sustainable. You should also get to grips with the different biological methods of extracting copper, and
the advantages in terms of sustainability of using these methods compared to traditional mining techniques.

Topic 10 — Using Resources

Life Cycle Assessments

Draw one line between each stage of a product's life and the correct example of that stage.

Life cycle stage	Example
Getting the Raw Materials	Coal being mined from the ground.
Manufacturing and Packaging	Plastic bags going on to landfill.
Using the Product	A car using fuel while driving.
Product Disposal	Books being made from wood pulp.

1 Life cycle assessments are carried out on products as a way of finding out what their environmental impact is.

Grade 6-7

1.1 State **two** potential impacts of a product on the environment when it is disposed of at the end of its useful life.

....

....

[2]

1.2 Suggest **one** environmental consideration of a product which can be easily quantified while carrying out a life cycle assessment.

....

[1]

1.3 Suggest why some pollutant effects are less straightforward to quantify.

....

[1]

1.4 If two different assessors carried out a life cycle assessment on identical products, would you expect the results to be exactly the same? Explain your answer.

....

....

....

[3]

1.5 Some life cycle assessments can be selective in assessing the effects of a product on the environment. Explain how and why selective life cycle assessments could be misused.

....

....

[2]

[Total 9 marks]

Potable Water

For each of the statements below circle whether the statement is **true** or **false**.

1) Potable water does not contain any dissolved substances.   True   or   False

2) Potable water is the same as drinking water.   True   or   False

3) Potable water can only be produced from fresh
   water found in rivers, streams and reservoirs.   True   or   False

1   This question is about potable water.   (Grade 4-6)

1.1   Which of the following is **not** a correct description of potable water? Tick **one** box.

Pure water   ☐

Water containing a small number of microbes   ☐

Water that is safe to drink   ☐

Water with a low concentration of salt   ☐

[1]

1.2   Fresh water is used to produce potable water. In the UK, where is the majority of fresh water
      sourced from during the production of potable water?

....
[1]

1.3   Draw **one** line between each treatment of water and the substances removed by the process.

| Passing water through filter beds | | Solid Waste |

| | | Microbes |

| Sterilisation | | Chemicals |

[2]

1.4   Name **three** things that can be used to sterilise fresh water.

....

....
[3]

[Total 7 marks]

2 Distillation can be used to desalinate sea water. **Figure 1** shows a set of equipment that could be used to desalinate sea water.

Figure 1

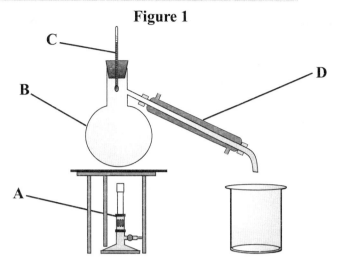

2.1 Name the components labelled **A** to **D** in the diagram.

A ....   **B** ....

C ....   **D** ....

[4]

2.2 Plan a method for using this equipment to produce a sample of potable water from seawater. You should include in your plan enough detail to allow someone to carry out the procedure safely and effectively.

....

....

....

....

....

[5]

2.3 Suggest a method, other than distillation, that could be used to desalinate water.

....

[1]

2.4 Explain why we do not use desalination to produce large quantities of potable water in the UK.

....

....

....

[2]

[Total 12 marks]

Exam Practice Tip

Some water costs a lot to make potable, some not so much. Make sure you understand the different processes that salty water and fresh water undergo to make it safe to drink and why the different processes are chosen.

Waste Water Treatment

1   Waste water must be treated before being reused or released into the environment. **Grade 4-6**

1.1   Which **two** of the following pollutants must be removed
from sewage and agricultural waste water?

Calcium ions, Ca^{2+} ☐

Organic matter ☐

Harmful microbes ☐

Sodium ions, Na^+ ☐

[2]

1.2   Industrial waste water must have further treatment compared to
sewage and agricultural waste water. Suggest why this.

....

[1]

[Total 3 marks]

2   This question is about the treatment of waste water in the form of sewage. **Grade 4-6**
Figure 1 shows a schematic diagram of a sewage treatment facility.

Figure 1

Waste Water → Screening → Sedimentation

B↗ aerobic digestion → released into rivers

A↘ X → natural gas → fertiliser

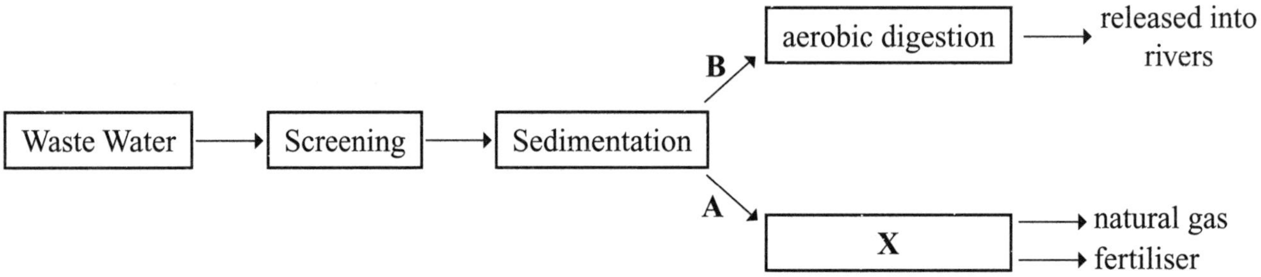

2.1   What is the purpose of the stage described as 'screening'?

....

....

[2]

2.2   What are the names given to the two substances produced by sedimentation?

Substance **A**: ....

Substance **B**: ....

[2]

2.3   What is the name of process **X**?

....

[1]

[Total 5 marks]

Topic 10 — Using Resources

The Haber Process

1 The Haber process is an important chemical process. **(Grade 4-6)**

1.1 Which two of these substances are used as reactants in the Haber process? Tick **two** boxes.

☐ Carbon ☐ Hydrogen ☐ Ammonia ☐ Oxygen ☐ Nitrogen

[1]

1.2 What is the product of the Haber process?

....

[1]

1.3 Why is the Haber process important to agriculture?

....

[1]

[Total 3 marks]

2 The Haber process is a reversible reaction usually conducted at a temperature of 450°C and a pressure of 200 atmospheres. The forward reaction is exothermic. **(Grade 7-9)**

2.1 Based only on Le Chatelier's principle, predict whether the reaction should be performed at high or low temperature to achieve maximum yield. Explain your answer.

....

....

....

[3]

2.2 Explain why the temperature that most Haber plants operate at is different to those conditions predicted as ideal using Le Chatelier's principle.

....

....

[1]

2.3 During the forward reaction of the Haber process, 4 moles of gas react to form 2 moles of gaseous product. Predict whether increasing the pressure of the gases in the Haber process increases or decreases the yield of product. Explain your answer.

....

....

....

[3]

2.4 Name **one** other factor, apart from the yield, that must be taken into consideration when choosing the temperature and pressure to operate the Haber process plant at?

....

[1]

[Total 8 marks]

NPK Fertilisers

1 This question is about some of the raw materials that are used in the production of NPK fertilisers. *(Grade 4-6)*

1.1 Name **two** compounds of potassium that are used to make NPK fertilisers.

....

[2]

1.2 How are these compounds obtained from natural deposits in the Earth?

....

[1]

[Total 3 marks]

2 This question is about the chemical processing of phosphate rock that is part of the production of NPK fertilisers. **Figure 1** shows one of the stages involved in the process used to convert phosphate rock into chemicals that are present in NPK fertilisers. *(Grade 6-7)*

Figure 1

$$\text{Phosphate rock} \xrightarrow{\text{Nitric acid}} \begin{array}{c} \text{Phosphoric acid} \\ + \\ \text{Substance } \mathbf{A} \end{array}$$

2.1 Name substance **A**.

....

[1]

2.2 Phosphate rock can also be reacted with sulfuric acid and phosphoric acid.
In each case, state the product(s) of the reaction.

Phosphate rock + sulfuric acid: ....

Phosphate rock + phosphoric acid: ....

[3]

2.3 The reactions of phosphate rock to produce fertilisers are carried out differently in industry to in the lab. Suggest a reason for each of the following differences:

The reaction in industry is carried out at much higher concentrations than in the lab.

....

....

[1]

In the lab, the product can be separated from the solution by crystallisation, but this technique isn't used in industry.

....

....

[1]

[Total 6 marks]

Topic 10 — Using Resources

Mixed Questions

1 A substance can be classified as an element, a compound or a mixture. *(Grade 4-6)*

1.1 Draw a line to connect each type of substance with an example of it.

compound		bromine water
element		helium
mixture		iron oxide

[2]

1.2 Formulations are a special type of mixture. Explain what is meant by the term 'formulation'.

....

....

[2]

[Total 4 marks]

2 Salt is the common name for sodium chloride. Salt can be obtained by mining rock salt, which also contains sand and insoluble bits of rock. *(Grade 4-6)*

2.1 How could the sand and bits of rock be separated from the salt? Tick **one** box.

Dissolve the rock salt in water and distill the water. ☐

Dissolve the rock salt in water and crystallise the sodium chloride. ☐

Dissolve the rock salt in water and filter. ☐

Dissolve the rock salt in water and carry out chromatography. ☐

[1]

2.2 Explain why sodium chloride is classified as a compound.

....

....

[2]

2.3 The structure of sodium chloride is shown in **Figure 1**.

Figure 1

● $= Cl^-$

○ $= Na^+$

State the type of bonding in sodium chloride.

....

[1]

[Total 4 marks]

3 Oxygen atoms have the electronic structure 2,6. (Grade 4-6)

3.1 State which group of the periodic table oxygen is in.
Explain your answer with reference to the electronic structure of oxygen.

Group: ....

Explanation: ....

[2]

3.2 Oxygen can react to form oxide ions. Predict, with reasoning the charge on an oxide ion.

....

....

[2]

3.3 When magnesium reacts with oxygen, it forms magnesium oxide. What type of reaction does magnesium undergo? Tick **one** box.

Displacement ☐   Oxidation ☐

Polymerisation ☐   Reduction ☐

[1]

[Total 5 marks]

4 Chlorine is a Group 7 element that exists as molecules of Cl_2. (Grade 4-6)

4.1 Complete the dot-and-cross diagram below to show the bonding in Cl_2.
You only need to show the outer electron shells.

Cl   Cl

[2]

4.2 Chlorine has two main isotopes — ^{35}Cl and ^{37}Cl. Explain the term 'isotope'.

....

....

[2]

4.3 Describe a test you could carry out for chlorine.   **PRACTICAL**
Include any observations you would expect.

....

....

[2]

4.4 Predict what happens if you mix chlorine water and sodium iodide solution. Explain your answer.

....

....

[2]

[Total 8 marks]

5 When sodium hydrogencarbonate reacts with ethanoic acid, the temperature of the surroundings decreases.

(Grade 4-6)

5.1 Is this reaction endothermic or exothermic?

....

[1]

5.2 Will the energy of the products be higher or lower than the energy of the reactants?

....

[1]

5.3 Explain the temperature change of this reaction in terms of the energy required to break the bonds in the reactants and the energy released when bonds are formed in the products.

....

....

[2]

5.4 Suggest one practical use of this reaction.

....

[1]

[Total 5 marks]

6 Pentane, C_5H_{12}, and decane, $C_{10}H_{22}$, are both hydrocarbons in the same homologous series.

(Grade 6-7)

6.1 What homologous series do pentane and decane belong to?

....

[1]

6.2 Name a process that could be used to separate pentane from decane.

....

[1]

6.3 Name a process that could be used to produce pentane from decane.

....

[1]

6.4 Predict, with reasoning, whether pentane or decane will have a higher boiling point.

....

....

....

[3]

6.5 Write a balanced symbol equation for the complete combustion of pentane in oxygen to form carbon dioxide and water.

....

[2]

[Total 8 marks]

7 Iron and silver are both metals. Iron has a melting point of 1538 °C, whilst silver has a melting point of 962 °C. Iron boils at 2862 °C and silver boils at 2162 °C.

Grade 6-7

7.1 Describe the bonding in metals, like iron and silver, when they are solid.

....

....

....

[3]

7.2 What states would iron and silver be in if they were both at a temperature of 1000 °C?

Iron: ....

Silver: ....

[2]

7.3 Use the information above to predict which metal has the strongest bonds. Explain your answer.

....

....

[3]

[Total 8 marks]

8 A student reacts lithium in an excess of water to produce lithium hydroxide and hydrogen gas. The equation for this reaction is: $2Li + 2H_2O \rightarrow 2LiOH + H_2$ Atomic masses, A_r: Li = 7, O = 16, H = 1

Grade 6-7

8.1 Lithium is the limiting reactant in this experiment. Explain the term 'limiting reactant'.

....

[1]

8.2 Calculate the relative formula mass of lithium hydroxide.

Relative formula mass = ...

[1]

8.3 A student performed this reaction using 1.75 g of lithium. How many moles is this?

Number of moles = ... mol

[2]

8.4 Use the symbol equation to calculate the mass, in g, of lithium hydroxide that would be produced using 0.50 mol lithium and an excess of water.

Mass = ... g

[3]

[Total 7 marks]

Mixed Questions

9 Models of the structure of the atom went through a lot of development in the early 1900s. (Grade 6-7)

9.1 Ernest Rutherford set up an experiment which involved firing alpha particles at a piece of gold foil. The results of the experiment did not match the prediction of what should have happened if the plum pudding model had been correct. Describe the model that Rutherford developed to replace the plum pudding model.

....

....

[2]

9.2 Describe the model of the atom used today, with reference to protons, neutrons and electrons.

....

....

....

[3]

[Total 5 marks]

PRACTICAL

10 When the following reaction is carried out in an unsealed reaction flask, the mass changes over the course of the reaction: $2HCl_{(aq)} + Na_2CO_{3(aq)} \rightarrow 2NaCl_{(aq)} + H_2O_{(l)} + CO_{2(g)}$ (Grade 6-7)

10.1 Explain, in terms of the particle model, why the mass of the reaction flask changes in the reaction.

....

....

....

[3]

10.2 The volume of gas produced by the reaction mixture can be used to investigate how the concentration of acid affects the rate of the reaction. Write a method the student could use to carry out this experiment.

....

....

....

....

....

....

[6]

10.3 At the start of an experiment, the gas syringe was empty. After 30 s, it contained 12.0 cm³ of gas. Calculate the mean rate of reaction during this time.

Mean rate = ... cm³/s

[2]

[Total 11 marks]

Mixed Questions

11 Carbon dioxide is a greenhouse gas found naturally in Earth's atmosphere. (Grade 6-7)

11.1 Millions of years ago, the amount of carbon dioxide in the atmosphere was much higher than it is today. State **two** ways in which carbon dioxide was removed from the atmosphere.

....

....

[2]

11.2 Human activities are increasing the amount of carbon dioxide in the atmosphere. Name **one** other greenhouse gas that is increasing due to human activity. Give an activity causing this increase.

....

....

[2]

11.3*Give examples of pollutant gases, other than carbon dioxide and the one mentioned in your answer to 11.2, that are produced by human activity. Explain how these gases form and how they impact on the environment and human health.

....

....

....

....

....

....

[6]

[Total 10 marks]

12 A cell was set up as shown in **Figure 2** to determine the relative reactivities of some metals. (Grade 6-7)

Figure 2

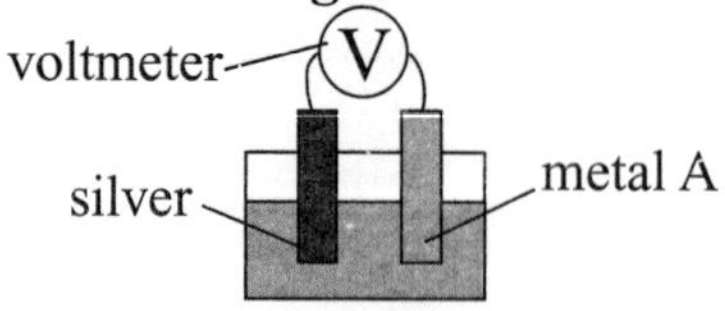

12.1 Given that silver is the least reactive of the metals, use the results in **Table 1** to put copper, iron and tin in order from most to least reactive.

Table 1

Metal A	Copper	Iron	Tin
Voltage (V)	0.46	1.24	0.94

....

[2]

12.2 Write a balanced equation, including state symbols, for the reaction that would occur if copper were mixed with silver chloride (AgCl) solution.

....

[3]

[Total 5 marks]

Mixed Questions

13 Copper is a metal found in a number of everyday mixtures and compounds. (Grade 6-7)

13.1 State, with reasoning, how you would extract copper from copper oxide.

....

....

[2]

13.2 Describe how bioleaching can be used to extract copper from low grade ores.

....

....

....

[3]

13.3 Copper can be drawn out into thin wires which are used in electrical cables.
Explain **two** ways in which the bonding and structure of copper make it suitable for this.

....

....

[2]

13.4 Statues can be made from bronze, an alloy of copper that contains tin. Explain how the structure of bronze makes it more suitable than pure copper for this use.

....

....

....

[3]

[Total 10 marks]

14 **Table 2** shows the properties of some polymers. (Grade 6-7)

Table 2

Polymer	Type of monomer(s)	Response when heated	Hardness
A	alkene	softens	flexible
B	alkene	does not soften	rigid
C	carboxylic acid and alcohol	softens	rigid

14.1 Explain what the difference in the response to heating tells you about the bonding in **A** and **B**.

....

....

[2]

14.2 Explain which polymer would be most suitable for making a plastic mug for hot drinks.

....

....

....

[3]

[Total 5 marks]

Mixed Questions

120

15 This question is about the compound ethanol, C_2H_5OH

15.1 Draw the displayed formula of ethanol.

[2]

15.2 Describe two methods, including the conditions, that can be used to produce ethanol.

....

....

....

....

....

[6]

[Total 8 marks]

16 Iron(II) sulfate, $FeSO_4$, is a compound of iron. (Grade 7-9) **PRACTICAL**

16.1 Describe a test you could carry out to confirm the identity of the cation. State any observations you would expect. Include a balanced symbol equation, with state symbols, for the reaction.

....

....

....

....

[5]

16.2 Describe a test you could carry out to confirm the identity of the anion. State any observations you would expect. Include a balanced symbol equation, with state symbols, for the reaction.

....

....

....

....

....

[5]

16.3 Describe a method to make pure, dry iron(II) sulfate crystals from iron(II) oxide and sulfuric acid.

....

....

....

....

[4]

[Total 14 marks]

Mixed Questions

17 Hydrogen gas can be produced by reacting carbon with steam. The symbol equation for the reaction is: $C + 2H_2O \rightarrow 2H_2 + CO_2$. Atomic masses, A_r: C = 12, H = 1, O = 16 **Grade 7-9**

17.1 Calculate the theoretical yield of hydrogen that would be produced from 24 g of carbon.

Theoretical yield = ... g

[4]

17.2 Calculate the volume of the theoretical yield of hydrogen gas, at room temperature and pressure.

Volume = ... dm^3

[1]

17.3 The actual yield of hydrogen was 4.8 g. Calculate the percentage yield of the reaction.

Percentage yield = ... %

[2]

[Total 7 marks]

18 Ammonia is made industrially using the following reaction: $N_{2(g)} + 3H_{2(g)} \rightleftharpoons 2NH_{3(g)}$ **Grade 7-9**
The reaction is carried out at 200 °C. The forward reaction is exothermic.

18.1 State the atom economy of this reaction. Give a reason for your answer.

....

....

[2]

18.2 State **one** reason why the percentage yield for this reaction will never be 100%.

....

[1]

18.3 Explain why the temperature chosen to carry out the reaction is a compromise.

....

....

....

....

....

[4]

[Total 7 marks]

Mixed Questions

PRACTICAL

19   Sulfuric acid, H_2SO_4, is a strong acid. In a titration experiment,
      a 0.10 mol/dm^3 solution of sulfuric acid was used to neutralise 25 cm^3 of
      a solution of sodium hydroxide, NaOH. The results are shown in **Table 3**.

Table 3

Titre	Rough	1	2	3
Volume (cm^3)	20.35	20.05	19.95	20.00

19.1 Write a balanced ionic equation for the reaction between sulfuric acid and sodium hydroxide.

....
                                                                                                                      [1]

19.2 Name the salt produced in this reaction.

....
                                                                                                                      [1]

19.3 Explain why Universal indicator shouldn't be used to find the endpoint of the titration, and
      suggest what could be used instead.

....

....
                                                                                                                      [2]

19.4 Calculate the number of moles of sulfuric acid in the rough titre.

                                                                          moles = ... g
                                                                                                                      [3]

19.5 Using the data in **Table 3**, calculate the mean titre volume.

                                                                     mean titre = ... cm^3
                                                                                                                      [2]

19.6 Calculate the concentration of the sodium hydroxide solution in mol/dm^3.

                                                                 concentration = ... mol/dm^3
                                                                                                                      [5]

                                                                                              [Total 14 marks]

Mixed Questions

20 Aluminium and iron are metals that corrode in oxygen and water. (Grade 7-9)

20.1 Describe how aluminium is extracted from its ore, Al_2O_3, using electrolysis.
You should include half equations for the reactions that occur at the anode and the cathode.

....

....

....

....

....

[6]

20.2 Explain why a block of aluminium would not be destroyed by corrosion.

....

....

[2]

20.3 Explain what galvanising is, and how it is used to protect iron from corrosion.

....

....

....

[3]

[Total 11 marks]

21 Carbon forms a large variety of different compounds. (Grade 7-9)

21.1 Put the carbon-based compounds butane, diamond and poly(propene) in order of melting point, from highest to lowest. Explain your answer.

Order: ....

Explanation: ....

....

....

[4]

21.2 Carbon nanotubes are a type of nanoparticle. Explain what a nanoparticle is, and why the general properties of nanoparticles may cause nanotubes to have different properties to bulk carbon.

....

....

[2]

21.3 Describe why nanotubes could be used in medicine, and give one possible risk of this application.

....

....

[2]

[Total 8 marks]

Mixed Questions

22 When zinc sulfate solution is electrolysed, the reaction at the cathode is: $2H^+_{(aq)} + 2e^- \rightarrow H_{2(g)}$ (Grade 7-9)

22.1 Explain why it is easier to discharge hydrogen rather than zinc, and what this tells you about the relative abilities of hydrogen and zinc to form positive ions.

....

....

[2]

22.2 What type of reaction takes place at the cathode? Explain your answer.

....

....

[2]

22.3 Write a balanced half equation for the reaction that occurs at the anode.

....

[2]

[Total 6 marks]

23 Some of the properties and reactions of different Group 1 metals are shown in **Table 4**. (Grade 7-9)

Table 4

Metal	Melting point (°C)	Typical ion that forms in reactions	Reaction with acid
Lithium	181	+1	vigorous
Sodium		+1	very vigorous, ignites
Potassium	63	+1	extremely vigorous, ignites

23.1 Predict **one** similarity and **one** difference between the properties of lithium shown in **Table 4** and cobalt, a typical transition metal.

....

....

[2]

23.2*Describe the similarities and differences in how lithium and sodium react, as shown in **Table 4**. Give reasons for them, based on the electronic structures of the atoms.

....

....

....

....

....

....

[6]

23.3 Using the data in the table, predict the melting point of sodium.

melting point = ... °C

[1]

[Total 9 marks]

Mixed Questions

Answers

Topic 1 — Atomic Structure and the Periodic Table

Page 1 — Atoms

Warm-up

The radius of an atom is approximately **0.1** nanometres. The radius of the nucleus is around 1×10^{-14} metres. That's about **1/10 000** of the radius of an atom.

An atom doesn't have an overall **charge** as it has equal numbers of **protons/electrons** and **electrons/protons**.

1.1 nucleus *[1 mark]*
1.2 −1 *[1 mark]*
1.3 neutron: 0 charge *[1 mark]*
    proton: +1 charge *[1 mark]*
2.1 mass number = 39 *[1 mark]*
2.2 atomic number = 19 *[1 mark]*
2.3 protons = 19 *[1 mark]*
    neutrons = mass number − atomic number
    　　　= 39 − 19 = **20** *[1 mark]*
    electrons = 19 *[1 mark]*

Page 2 — Elements

1.1 Atoms are the smallest part of an element that can exist *[1 mark]*.
1.2 They have the same number of protons / 17 protons *[1 mark]* but a different number of neutrons / $^{35}_{17}Cl$ has 2 less neutrons than $^{37}_{17}Cl$ *[1 mark]*.

2.1

Isotope	No. of Protons	No. of Neutrons	No. of Electrons
^{32}S	16	16	16
^{33}S	16	17	16
^{34}S	16	18	16
^{36}S	16	20	16

[3 marks — 1 mark for each correct column]

2.2 Relative atomic mass = $[(94.99 \times 32) + (0.75 \times 33) + (4.25 \times 34) + (0.01 \times 36)] \div (94.99 + 0.75 + 4.25 + 0.01)$
    = 3209.29 ÷ 100 = 32.0929 = **32.1** *[2 marks for correct answer, otherwise one mark for using correct equation]*
2.3 X and Z are isotopes *[1 mark]*. They have the same atomic number / same number of protons *[1 mark]* but different mass numbers / number of neutrons *[1 mark]*.

Page 3 — Compounds

1.1 It contains two elements chemically combined *[1 mark]*.
1.2 4 *[1 mark]*
A molecule of ammonia contains 1 nitrogen atom and 3 hydrogen atoms making a total of 4 atoms altogether.
2.1 sodium chloride *[1 mark]*
2.2 Any one of: **B.** NaCl / **C.** C_2H_4 / **E.** H_2O *[1 mark]*
    It contains two or more elements chemically combined (in fixed proportions) *[1 mark]*.
2.3 6 *[1 mark]*
C_2H_4 contains 2 carbon atoms and 4 hydrogen atoms.
2.4 Yes, a new compound has been made as the atoms in C_2H_6 are in different proportions to the atoms in C or F / there are a different number of hydrogen atoms in the molecule *[1 mark]*.

Page 4 — Chemical Equations

Warm-up
1 True
2 False
3 True
4 True
1.1 sodium + chlorine → sodium chloride *[1 mark]*
1.2 $2Na + Cl_2 \rightarrow 2NaCl$ *[1 mark]*
2.1 $4NH_3 + 5O_2 \rightarrow 4NO + 6H_2O$ /
    $2NH_3 + 2.5O_2 \rightarrow 2NO + 3H_2O$ *[1 mark]*
2.2 E.g. there are 7 oxygen atoms on the left hand side of the equation and only 6 on the right hand side *[1 mark]*.

Page 5 — Mixtures and Chromatography

1.1 Mixture *[1 mark]*. Air consists of two or more elements or compounds *[1 mark]* that aren't chemically combined together *[1 mark]*.
1.2 No *[1 mark]*, as argon is an element in a mixture. Chemical properties are not affected by being in a mixture *[1 mark]*.
2 How to grade your answer:
    Level 0: Nothing written worthy of credit *[No marks]*.
    Level 1: Some explanation or description given but little detail and key information missing *[1–2 marks]*.
    Level 2: Clear description of method and some explanation of results but some detail missing *[3–4 marks]*.
    Level 3: A clear and detailed description of method and a full explanation of results *[5–6 marks]*.
    Here are some points your answer may include:
    <u>Setting up the experiment</u>
    Draw a line in pencil near the bottom of a piece of chromatography paper.
    Place a small sample of each ink on the pencil line.
    Pour a shallow layer of water / solvent into a beaker.
    Place the chromatography paper in the container.
    The water should be below the pencil line and the ink spots.
    Place a lid on the container and wait for the solvent to rise to near the top of the paper.
    Remove the paper from the container when the solvent has risen close to the top of the paper.
    <u>Explanation of results</u>
    A shows one spot, so only contains one dye.
    B shows two spots that have separated, so contains two dyes.
    C shows three spots that have separated, so contains three dyes.
    B and C are mixtures as they contain more than one element or compound not chemically combined together.
    B and C contain at least one of the same dyes.

Page 6 — More Separation Techniques

1.1 Add water to the mixture to dissolve the potassium chloride *[1 mark]*. Filter the mixture. The chalk will stay on the filter paper, *[1 mark]* the dissolved potassium chloride will pass through *[1 mark]*.
1.2 E.g. evaporate the potassium chloride solution to a much smaller volume and then leave it to cool *[1 mark]*.
2.1 Add the mixture to methylbenzene. The sulfur will dissolve (the iron will not dissolve) *[1 mark]*. Filter the solution to obtain the insoluble iron *[1 mark]*. Evaporate the methylbenzene to obtain crystals of sulfur *[1 mark]*.
2.2 No, the student is incorrect *[1 mark]*. The iron and sulfur are chemically combined in iron(II) sulfide / iron(II) sulfide is a compound *[1 mark]* so chemical methods would be needed to separate them out *[1 mark]*.

Page 7 — Distillation

1   Simple distillation *[1 mark]*

2.1   Place a stopper / stopper with a thermometer in the top of the distillation flask *[1 mark]*.

2.2   The solution is heated/boiled and the aspirin evaporates first as it has a lower boiling point than the impurity *[1 mark]*. There is cold water flowing through the (Liebig) condenser *[1 mark]*. This condenses the gaseous aspirin back into a liquid which is then collected *[1 mark]*.

2.3   The aspirin has a boiling point greater than 100 °C / greater than the boiling point of water *[1 mark]*. So it would not evaporate *[1 mark]*.

Pages 8-9 — The History of The Atom

Warm-up

     New experimental evidence can disprove models — **True**

     Scientific models can be based on existing theories and new experimental evidence — **True**

     Older scientific theories must be ignored when new ones are adopted — **False**

1.1   Tiny solid spheres that can't be divided *[1 mark]*.

1.2   Plum pudding model — A positively charged 'ball' with negatively charged electrons in it *[1 mark]*.

     Bohr's model — Electrons in fixed orbits surrounding a small positively charged nucleus *[1 mark]*.

     Rutherford's nuclear model — A small positively charged nucleus surrounded by a 'cloud' of negative electrons *[1 mark]*.

1.3   neutron *[1 mark]*

2.1   Most of the atom is "empty" space *[1 mark]*.

2.2   Niels Bohr *[1 mark]*

3.1   Atoms are neutral / have no overall charge *[1 mark]*. Therefore there must have been positive charge to balance the negative charge of the electrons *[1 mark]*.

3.2   How to grade your answer:

     Level 0:   Nothing written worthy of credit *[No marks]*.

     Level 1:   A brief description of either the nuclear or the 'plum pudding' model is given *[1 to 2 marks]*.

     Level 2:   A description of both the nuclear model and the plum pudding model is given and some comparisons made *[3 to 4 marks]*.

     Level 3:   A full comparison of the models is given and similarities and differences are clearly explained *[5 to 6 marks]*.

     Here are some points your answer may include:

     <u>Similarities</u>

     They both have areas of positive charge.

     They both have electrons.

     They are both neutral overall.

     <u>Differences</u>

     Positive charge isn't divided into protons in plum pudding model.

     Plum pudding model does not have a nucleus but has a 'ball' of positive charge instead.

     Plum pudding model does not have neutrons or protons, it only has electrons surrounded by a positive charge.

     Plum pudding model does not have shells of electrons (surrounding nucleus), the electrons are arranged randomly within a sphere of positive charge.

     Modern nuclear model has most of the mass concentrated in the nucleus but the plum pudding model has the mass spread evenly throughout the entire atom.

Page 10 — Electronic Structure

1.1   2,8,8,2 *[1 mark]*

1.2   The electrons in an atom occupy the lowest energy levels/ innermost shell first *[1 mark]*. The innermost shell/lowest energy level can hold 2 electrons *[1 mark]*.

2.1   Chlorine: 2,8,7 *[1 mark]*

2.2

     [1 mark for correct number of electrons, 1 mark for correct arrangement]

You don't have to have the electrons paired up on the diagram. As long as there is the same number of electrons on the same shells you get the marks.

2.3   Phosphorus/P *[1 mark]*

Page 11 — Development of The Periodic Table

1.1   He left gaps so that elements with similar properties were in the same group / for elements that had not yet been discovered *[1 mark]*.

1.2   **D**. Between 2.4 and 7.2 g/cm^3 *[1 mark]*. **E**. EkO$_2$ *[1 mark]* **F**. EkCl$_4$ *[1 mark]* **G**. Very slow *[1 mark]*.

2.1   Protons (neutrons and electrons) had not been discovered / atomic numbers weren't known *[1 mark]*.

2.2   Ar and K / Te and I *[1 mark]*.

2.3   Isotopes of an element have different numbers of neutrons/ different atomic masses *[1 mark]*, but the same chemical properties *[1 mark]*.

Page 12 — The Modern Periodic Table

1.1   By atomic number / proton number *[1 mark]*.

1.2   Similar properties occur at regular intervals / there are repeating patterns in the properties of the elements *[1 mark]*.

1.3   They have the same number of outer shell electrons *[1 mark]*.

2.1   Group 2 *[1 mark]*. The atom has 2 outer shell electrons. *[1 mark]*.

2.2   Period 3 *[1 mark]*. The atom has 3 shells of electrons *[1 mark]*.

2.3   Magnesium/Mg *[1 mark]*

2.4   Choose one from: beryllium / calcium / strontium / barium / radium *[1 mark]*

Page 13 — Metals and Non-Metals

1.1   Metals: Towards the left and bottom. Non-metals: Towards the right and top *[1 mark]*.

1.2   Elements that react to form positive ions are metals *[1 mark]*.

1.3   Any one from: e.g. good electrical conductor / good thermal conductor / strong / high boiling point / high melting point / malleable *[1 mark]*.

1.4   Both are metals that lose their (2 or 3) outer shell electrons *[1 mark]* to form positive ions *[1 mark]*.

2.1   Iron is a metal and therefore strong / good at conducting heat/electricity / has a high boiling/melting point *[1 mark for each property up to a maximum of 2 marks]*. Sulfur is a non-metal and therefore is more brittle / has a lower density / doesn't conduct electricity / dull looking *[1 mark for each property up to a maximum of 2 marks]*.

2.2   E.g. they can form more than one ion *[1 mark]*, they make good catalysts *[1 mark]*.

2.3   Any two from: e.g. cobalt / copper / manganese / nickel / chromium / vanadium *[1 mark for each correct answer]*

Any two transition metals (apart from iron) will get you the marks.

Page 14 — Group 1 Elements

1.1 **Y** *[1 mark]*. As element **Y** has a higher melting point, it must be higher up the group than **X** *[1 mark]*.
The higher up the group an element is, the lower its atomic number.

1.2 $2X_{(s)} + 2H_2O_{(l)} \rightarrow 2XOH_{(aq)} + H_{2(g)}$
[1 mark for correct reactants and products and 1 mark for balanced equation. Half the ratio is acceptable]

1.3 Anything between 8-14 *[1 mark]*.

2.1

	Boiling Point / °C	Radius of atom / pm
Rb	687.8	248
Cs	670.8	265
Fr	Accept lower than 670.8	Accept greater than 265

[1 mark for each correct answer]

2.2 Francium would be more reactive than caesium *[1 mark]*. As you go further down the group the outer electron is further away from the nucleus *[1 mark]*, so the attraction between the nucleus and the electron decreases and the electron is more easily lost *[1 mark]*.

2.3 Formula: Fr_3P *[1 mark]*
Equation: $12Fr + P_4 \rightarrow 4Fr_3P$ *[1 mark for correct reactants and products, 1 mark for correctly balancing the equation]*

Pages 15-16 — Group 7 Elements

Warm-up

Fluorine
Chlorine
Bromine
Iodine

1.1 They are non-metals that exist as molecules of two atoms *[1 mark]*.

1.2 Chlorine is more reactive than bromine *[1 mark]*. This is because chlorine's outer shell is closer to the nucleus *[1 mark]* so it's easier for chlorine to gain an electron when it reacts *[1 mark]*.

Because of the increasing distance between the nucleus and the outer shell, reactivity decreases down the group. Bromine is further down the group than chlorine, it's outer shell is further away from the nucleus and therefore it's less reactive than chlorine.

1.3 P *[1 mark]*

2.1 $2Fe + 3Br_2 \rightarrow 2FeBr_3$ *[1 mark for Br_2 and 1 mark for balanced equation. Half the ratio is acceptable]*

2.2 −1 *[1 mark]*
All halide ions form ions with a −1 charge.

3.1 chlorine + potassium bromide → **potassium chloride** + bromine *[1 mark]*

3.2 The solution will turn orange *[1 mark]*.

3.3 displacement *[1 mark]*

3.4 No *[1 mark]*, as chlorine is less reactive than fluorine *[1 mark]*.

4.1 The halogens have seven electrons in their outer shell *[1 mark]*. As you go further down the group additional shells are added so the outer electron is further away from the nucleus *[1 mark]*.

4.2 Astatine will react more slowly than fluorine *[1 mark]* since reactivity decreases down the group *[1 mark]*. Both astatine and fluorine have 7 outer shell electrons so react in a similar way *[1 mark]*. So astatine will react with hydrogen to form hydrogen astatide/HAt *[1 mark]*. $H_2 + At_2 \rightarrow 2HAt$ *[1 mark]*

Page 17 — Group 0 Elements

1.1 Rn Boiling Point: Above −108 °C *[1 mark]*, Xe Density: Between 0.0037 and 0.0097 *[1 mark]*, Ar Atomic Radius: Less than 109 pm *[1 mark]*.

1.2 Krypton is unreactive *[1 mark]*. It has a stable electron arrangement / full outer shell / 8 electrons in its outer shell *[1 mark]*.

1.3 Helium only has 2 electrons in its outer shell. The rest of the noble gases have 8 *[1 mark]*.

2.1 Noble gases are unreactive / they have stable electron arrangements / full outer shells / 8 electrons in their outer shell *[1 mark]*.

2.2 Iodine is much less reactive than fluorine *[1 mark]*.

2.3 Neon solidified at −249 °C and xenon at −112 °C *[1 mark]* Boiling points increase down the group *[1 mark]* and xenon is further down the group than neon so will have the higher boiling point *[1 mark]*.

Topic 2 — Bonding, Structure and Properties of Matter

Page 18 — Formation of Ions

1.1 Metal atoms usually lose electrons to become positive ions *[1 mark]*.

1.2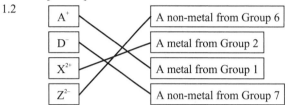

[2 marks if all four correct, otherwise 1 mark if two correct]

2.1 2− *[1 mark]*

2.2 2,8,8 *[1 mark]*. Sulfur gains two electrons *[1 mark]* to achieve a noble gas electronic structure/a full outer shell *[1 mark]*.

2.3 Argon/Ar *[1 mark]*

Pages 19-20 — Ionic Bonding

Warm-up

Dot and cross diagram	Ionic formula
$\left[Na\right]^+ \left[Cl\right]^-$	**NaCl**
$\left[Na\right]^+ \left[O\right]^{2-} \left[Na\right]^+$	**Na₂O**
$\left[Cl\right]^- \left[Mg\right]^{2+} \left[Cl\right]^-$	**MgCl₂**

1.1 calcium chloride *[1 mark]* and potassium oxide *[1 mark]*
Compounds that contain ionic bonding have to be made up of a metal and a non-metal. All the other options only contain non-metals, so can't be held together by ionic bonds.

1.2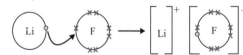

[1 mark for arrow showing electron transfer from Li to F, 1 mark for correct electronic structure of fluoride ion, with seven crosses and one dot, 1 mark for correct charges on the ions.]

1.3 electrostatic attraction / electrostatic force *[1 mark]*

1.4 E.g. the particles in the compound are oppositely charged ions / have opposite charges / the bond is formed by electrons being transferred from one atom to another *[1 mark]*.

2.1 $\left[Mg\right]^{2+}$ *[1 mark for no electrons in outer shell, 1 mark for correct charge]*

If you showed the second electron shell of magnesium containing eight electrons as dots, you also get the mark.

[1 mark for eight electrons in the outer shell, with two dots and six crosses, 1 mark for correct charge]

2.2   E.g. the magnesium atom transfers two electrons to the oxygen atom *[1 mark]*. A magnesium ion with a 2+ charge forms *[1 mark]*, and an oxide ion with a 2– charge forms *[1 mark]*. The oppositely charged ions are attracted to each other by electrostatic attraction *[1 mark]*.

3.1   Element X: Group 7 *[1 mark]*
Reason: Any one of, e.g. it has formed an ion by gaining 1 electron / it forms 1– ions / the uncharged element would have seven electrons in its outer shell *[1 mark]*.
Element Z: Group 2 *[1 mark]*
Reason: Any one of, e.g. it has formed an ion by losing 2 electrons / it forms 2+ ions / the uncharged element would have two electrons in its outer shell *[1 mark]*.

3.2   How to grade your answer:
Level 0:   There is no relevant information *[No marks]*.
Level 1:   The discussion is limited and doesn't mention both the uses and limitations of dot and cross diagrams *[1 to 2 marks]*.
Level 2:   There is some discussion of dot and cross diagrams, with at least one use and one limitation covered *[3 to 4 marks]*.
Level 3:   The discussion is comprehensive in evaluating both the uses and limitations of dot and cross diagrams *[5 to 6 marks]*.
Here are some points your answer may include:
Dot and cross diagrams show:
Charge of the ions.
The arrangement of electrons in an atom or ion.
Which atoms the electrons in an ion originally come from.
Empirical formula (correct ratio of ions).
Dot and cross diagrams do not:
Show the structure of the compound.
Correctly represent the sizes of ions.

Pages 21-22 — Ionic Compounds

Warm-up
In an ionic compound, the particles are held together by **strong** forces of attraction. These forces act **in all directions** which results in the particles bonding together to form **giant lattices**.

1.1   conduct electricity in the solid state *[1 mark]*
1.2   giant ionic lattice *[1 mark]*
2.1   Sodium chloride contains positive sodium ions (Na^+) *[1 mark]* and negative chloride ions (Cl^-) *[1 mark]* that are arranged in a regular lattice/giant ionic lattice *[1 mark]*. The oppositely charged ions are held together by electrostatic forces acting in all directions *[1 mark]*.
2.2   To melt sodium chloride, you have to overcome the very strong electrostatic forces/ionic bonds between the particles *[1 mark]*, which requires lots of energy *[1 mark]*.
3.1   E.g.

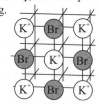

[1 mark for K$^+$ ions, 1 mark for Br$^-$ ions, 1 mark for correct structure, with alternating ions]
You'd also get the marks if you labelled all the white circles as Br$^-$ and all the grey circles as K$^+$.
3.2   Advantage: Any one of, e.g. the diagram shows the 3D arrangement of the ions / it suggests the structure is extended / it shows the regular (repeating) pattern of the ions *[1 mark]*.
Disadvantage: Any one of, e.g. the diagram doesn't correctly represent the sizes of ions / it shows gaps between the ions *[1 mark]*.

3.3   KBr *[1 mark]*
Remember that the overall charge of the ionic compound must be neutral. So you can work out the empirical formula by seeing that you only need one bromide ion to balance the charge on a potassium ion.
3.4   Boiling point: Potassium bromide has a giant structure with strong ionic bonds *[1 mark]*. In order to boil, these bonds need to be broken, which takes a lot of energy *[1 mark]*.
Electrical conductivity of solid: The ions are in fixed positions in the lattice *[1 mark]* and so are not able to move and carry a charge through the solid *[1 mark]*.
Electrical conductivity of solution: In solution, the ions are free to move *[1 mark]* and can carry a charge from place to place *[1 mark]*.

Pages 23-24 — Covalent Bonding

1.1   They share a pair of electrons *[1 mark]*.
1.2   Non-metals *[1 mark]*
1.3 BH_3 *[1 mark]*
Find the molecular formula by counting up how many atoms of each element there are in the diagram.
2

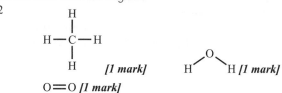

Each line represents one covalent bond. Oxygen has a double bond, so you need to draw two lines between the oxygen atoms to show this.
3.1   E.g. it contains only non-metals *[1 mark]* and Figure 1 shows shared electrons *[1 mark]*.
3.2   Any two from, e.g. they don't show how the atoms are arranged in space / they don't show the relative sizes of the atoms *[2 marks — 1 mark for each correct answer]*.
3.3   One electron from hydrogen and one from carbon form a shared pair *[1 mark]* that are attracted to the nuclei of the carbon and hydrogen atoms *[1 mark]* by electrostatic attraction *[1 mark]*.
4.1   Displayed formula: e.g. it shows how all the atoms in a molecule are connected in a simple way *[1 mark]*, but it doesn't show the 3D structure of the molecule / it doesn't show which atom the electrons in the bond originally come from *[1 mark]*.
Dot and cross diagram: e.g. it shows where the electrons in each covalent bond originally came from *[1 mark]* but it doesn't show the 3D structure of the molecule / they can become very complicated if the molecule is large *[1 mark]*.
3D model: e.g. it shows how all the atoms are arranged in space in relation to each other / it shows the correct bond angles in the molecule *[1 mark]* but it quickly becomes complicated for large molecules / you can't tell which atom in the bonds the electrons originally came from *[1 mark]*.
4.2   The displayed formula *[1 mark]* would be the best as it is easy to see how the atoms in a large molecule are connected without the diagram becoming too complicated *[1 mark]*.

Pages 25-26 — Simple Molecular Substances

1.1   The bonds between the atoms are strong *[1 mark]*, but the forces between the molecules are weak *[1 mark]*.
1.2   The weak forces between the molecules / the intermolecular forces *[1 mark]*.
2.1

[1 mark for correct number of electrons, 1 mark for one shared pair]
2.2
[1 mark for correct number of electrons, 1 mark for three shared pairs]

Answers

2.3 E.g. N_2 has a triple covalent bond, whilst HCl has a single covalent bond *[1 mark]*.

3.1 Simple molecular substances have weak forces between molecules *[1 mark]* so not much energy is needed to overcome them/they normally have low melting points *[1 mark]*.

3.2 Iodine won't conduct electricity *[1 mark]* because the I_2 molecules aren't charged / the electrons aren't free to move so can't carry a charge *[1 mark]*.

4.1 When methane boils, the forces between the molecules are overcome *[1 mark]* and it turns from a liquid into a gas *[1 mark]*. Methane is a smaller molecule then butane *[1 mark]* so the forces between the molecules are weaker *[1 mark]* and less energy is needed to overcome them *[1 mark]*.

4.2 Carbon needs four more electrons to get a full outer shell, and does this by forming four covalent bonds *[1 mark]*. Hydrogen only needs one more electron to complete its outer shell, so can only form one covalent bond *[1 mark]*.

Remember that the outer electron shell in hydrogen only needs two electrons to be filled, not eight like other electron shells.

4.3 Four *[1 mark]*. Silicon has four outer electrons so needs four more to get a full outer shell / silicon has the same number of outer shell electrons as carbon so will form the same number of bonds *[1 mark]*.

Page 27 — Polymers and Giant Covalent Substances

Warm-up

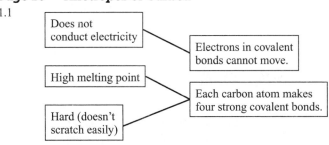

1.1 Ammonia *[1 mark]*

Ammonia has a simple covalent structure — it forms small molecules.

1.2 The covalent bonds are very strong *[1 mark]*, so a lot of energy is needed to break them *[1 mark]*.

2.1 $(C_2H_4)_n$ *[1 mark]*

2.2 Solid *[1 mark]*. The molecule is very large and so the intermolecular forces are strong *[1 mark]* and need lots of energy to be broken *[1 mark]*.

2.3 covalent bonds *[1 mark]*

Page 28 — Allotropes of Carbon

1.1

Does not conduct electricity — Electrons in covalent bonds cannot move.

High melting point ⟍

Hard (doesn't scratch easily) — Each carbon atom makes four strong covalent bonds.

[2 marks if all correct, otherwise 1 mark if one correct]

1.2 A: graphene *[1 mark]*
B: buckminster fullerene *[1 mark]*
C: carbon nanotube / fullerene *[1 mark]*

1.3 Any one of, e.g. to strengthen materials / to deliver drugs into the body / as a catalyst / as a lubricant / in electronics *[1 mark]*

2.1 Graphite is made up of sheets of carbon atoms arranged in hexagons *[1 mark]*, with weak forces between the sheets *[1 mark]*. Each carbon atom forms three covalent bonds *[1 mark]*, and has one delocalised electron *[1 mark]*.

2.2 Graphite has delocalised electrons *[1 mark]* which are free to move through the substance and carry an electric current *[1 mark]*.

Page 29 — Metallic Bonding

1.1 E.g.

Metal ions in a regular pattern

Delocalised electrons

[1 mark for regular arrangement of metal ions, 1 mark for delocalised electron, 1 mark for correct labels]

1.2 There is a strong electrostatic attraction *[1 mark]* between the delocalised electrons and the positive metal ions *[1 mark]*.

1.3 High *[1 mark]* because the bonding is strong so requires lots of energy to break *[1 mark]*.

1.4 Good *[1 mark]* because the electrons are free to move throughout the structure and carry an electrical charge *[1 mark]*.

2.1 Metallic structures have layers of atoms *[1 mark]* that are able to slide over one another *[1 mark]*.

2.2 Atoms of different elements are different sizes *[1 mark]*. Adding atoms of a different size to a pure metal distorts the layers *[1 mark]* making it harder for them to slide over one another *[1 mark]*.

Page 30 — States of Matter

1.1 solid, liquid, gas *[1 mark]*

1.2 $NaCl_{(s)}$: solid *[1 mark]*
$O_{2(g)}$: gas *[1 mark]*
$Hg_{(l)}$: liquid *[1 mark]*

2.1 solid spheres *[1 mark]*

2.2 liquid *[1 mark]*

2.3 Any two from: melting / boiling / condensing / freezing *[1 mark for each]*

2.4 Any two from: e.g. the model says nothing about forces between particles / particles aren't really spheres / particles are mostly empty space, not solid *[1 mark for each]*.

Page 31 — Changing State

1.1 melting *[1 mark]*

1.2 boiling point *[1 mark]*

1.3 The bonds are strong *[1 mark]*.

2.1 sodium chloride *[1 mark]*

At 900 °C, water would be a gas and copper would be a solid.

2.2 Sodium chloride *[1 mark]* and water *[1 mark]*.

At 1500 °C, copper would be a liquid.

2.3 Boiling sodium chloride *[1 mark]*.

2.4 No *[1 mark]*. When copper boils, the metallic bonds are broken *[1 mark]*, but when water boils only the intermolecular forces are broken *[1 mark]*, so you can't tell anything about the strength of the covalent bonds *[1 mark]*.

Pages 32-33 — Nanoparticles

Warm-up

1 True
2 False
3 True
4 True
5 False

1

The maximum size, in nanometres (nm), of a nanoparticle. — 100

An approximate number of atoms present in one nanoparticle. — 500

The possible size, in nanometres (nm), of a dust particle. — 5000

[2 marks if all three correct, otherwise 1 mark if one correct]

2.1   10 nm × 10 nm × 10 nm = **1000 nm³** *[1 mark]*

2.2   Area of one side = 10 nm × 10 nm = 100 nm²
A cube has six sides, so total surface area = 6 × 100 nm²
= **600 nm²** *[2 marks for correct answer, otherwise 1 mark for the area of one side]*

2.3   Surface area to volume ratio = 600 nm² ÷ 1000 nm³
= **0.6 nm⁻¹** *[1 mark]*

2.4   Volume = 1 nm × 1 nm × 1 nm = 1 nm³
Surface area of one side = 1 nm × 1 nm = 1 nm²
Total surface area of cube = 6 × 1 nm² = 6 nm²
Surface area to volume ratio = 6 nm² ÷ 1 nm³ = **6 nm⁻¹**
[4 marks for correct answer, otherwise 1 mark for correct volume, 1 mark for correct surface area of one side, 1 mark for correct total surface area]
E.g. decreasing the side length by a factor of 10 increases the surface area to volume ratio by a factor of 10 *[1 mark]*.

3.1   Surface area to volume ratio = 0.12 ÷ 10 = **0.012 nm⁻¹**
[1 mark]

The side length has increased by a factor of ten, so the surface area to volume ratio will decrease by a factor of 10.

3.2   Material Y *[1 mark]*, because the particles have a lower surface area to volume ratio *[1 mark]*.

Page 34 — Uses of Nanoparticles

1.1   E.g. suncreams containing nanoparticles give better skin coverage *[1 mark]* and are better at absorbing UV rays *[1 mark]*.

1.2   E.g. the nanoparticles in suncreams may affect people's health *[1 mark]*, and when they're washed away they could damage the environment *[1 mark]*.

2.1   Material: carbon nanotubes *[1 mark]*
Reason: e.g. the carbon nanotubes trap the drug molecules inside, and release them when they're at the right place in the body *[1 mark]*.

2.2   Material: silver *[1 mark]*
Reason: e.g. silver nanoparticles are antibacterial, so will help to kill any bacteria that are in the water *[1 mark]*.

2.3   Material: carbon nanotubes *[1 mark]*
Reason: e.g. carbon nanotubes are strong, so will strengthen the sports equipment, but they are also light, so won't add much weight to the equipment *[1 mark]*.

Topic 3 — Quantitative Chemistry

Page 35 — Relative Formula Mass

1
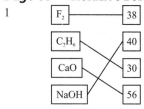
[2 marks if all four correct, otherwise 1 mark if two correct]

2.1 M_r(MgO) = 24 + 16 = 40 *[1 mark]*

percentage by mass of magnesium = $\dfrac{A_r(\text{Mg})}{M_r(\text{MgO})} \times 100$

= $\dfrac{24}{40} \times 100$ = **60%** *[1 mark]*

2.2   Mass of magnesium ions = $200 \times \dfrac{15}{100}$ = **30 g** *[1 mark]*

2.3   Mass of magnesium oxide containing 30 g of

magnesium ions = $30 \div \dfrac{60}{100}$ = **50 g** *[1 mark]*

If you used the percentage mass of magnesium ions as 40% and the mass of magnesium ions in the mixture as 20 g, your answer will also be 50 g.

Page 36 — The Mole

Warm-up:
    6.02×10^{23}

1.1 M_r of carbon dioxide = 12 + (16 × 2) = **44** *[1 mark]*

1.2   Moles of carbon dioxide = 110 ÷ 44 = **2.5 mol** *[1 mark]*

1.3   1 mole of carbon dioxide would weigh more *[1 mark]*.
It has a higher relative formula mass *[1 mark]*.

2.1   2 mol sulfur = 2 × 32 g = **64 g** *[1 mark]*

2.2 M_r of iron sulfide = 56 + 32 = 88
Moles of iron sulfide = 44 ÷ 88 = **0.50 mol** *[2 marks for correct answer, otherwise 1 mark for correct working]*

2.3   The number of atoms in 3 moles of sulfur is greater than the number of molecules in 2 moles of iron sulfide *[1 mark]*.
There's the same number of atoms in 1 mole of sulfur as there are molecules in 1 mole of iron sulfide so in 3 moles of sulfur there will be more atoms than there are molecules in 2 moles of iron sulfide *[1 mark]*.

Pages 37-38 — Conservation of Mass

1.1 $2Mg + O_2 \rightarrow 2MgO$ *[1 mark]*

1.2   Mass of oxygen = 20 g of MgO – 12 g of Mg = **8 g** *[2 marks for correct answer, otherwise 1 mark for correct working]*

2.1   The mass of reactants equals the mass of products in a chemical reaction *[1 mark]*. Atoms are not made or destroyed during a chemical reaction *[1 mark]*. So, there must be the same number of each type of atom in the products as in the reactants *[1 mark]*.

2.2   The mass of the powder would increase *[1 mark]*.
Oxygen gas was not included as part of the original measurement *[1 mark]*. Particles of oxygen are added to the zinc to form zinc oxide powder *[1 mark]*.

3.1   The measurement is correct *[1 mark]*. Carbon dioxide (a gas) is produced and released into the atmosphere *[1 mark]*. So, the student only measured the mass of the solid product, not both reactants *[1 mark]*.

3.2 M_r of sodium oxide = 106 – 44 = **62** *[1 mark]*

3.3   Moles of Na_2CO_3 = 53 ÷ 106 = 0.50
For every mole of Na_2CO_3 that reacts, 1 mole of CO_2 is produced. Only 0.50 moles of Na_2CO_3 react so 0.50 moles of CO_2 are produced.
Mass of carbon dioxide = 0.50 × 44 = **22 g** *[3 marks for correct answer, otherwise 1 mark for 0.50 moles of Na_2CO_3 and 1 mark for a 1:1 molar ratio]*

To work out a molar ratio, you need to use the balanced symbol equation for the reaction. The numbers in front of the chemical formulas show the number of moles of a substance that react or are produced in the reaction. In this question, for every 1 mole Na_2CO_3 heated, 1 mole of carbon dioxide is produced — a 1:1 molar ratio.

3.4   Mass of sodium oxide = 53 g – 22 g = **31 g** *[1 mark]*

Pages 39-40 — The Mole and Equations

Warm-up
    3

1     $H_2SO_4 + 2NaOH \rightarrow Na_2SO_4 + 2H_2O$ *[1 mark]*

2.1   Moles of sodium = 9.2 ÷ 23 = **0.4 mol** *[1 mark]*

2.2 M_r of water = (1 × 2) + 16 = 18
Moles of water = 7.2 g ÷ 18 = **0.4 mol** *[2 marks for correct answer, otherwise 1 mark for correct working]*

2.3   Divide the number of moles of each substance by the lowest of these number of moles (0.2 mol) to give the molar ratios.
Na = 0.4 ÷ 0.2 = 2 mol
H_2O = 0.4 ÷ 0.2 = 2 mol
NaOH = 0.4 ÷ 0.2 = 2 mol
H_2 = 0.2 ÷ 0.2 = 1 mol
$2Na + 2H_2O \rightarrow 2NaOH + H_2$ *[3 marks for correct answer, otherwise 1 mark for correct method and 1 mark for at least 2 correct numbers in the equation]*

3.1   Moles of methane = 8 g ÷ 16 = 0.5 mol
        Moles of oxygen = 32 g ÷ 32 = 1 mol
        Moles of carbon dioxide = 22 g ÷ 44 = 0.5 mol
        Moles of water = 18 g ÷ 18 = 1 mol *[1 mark]*
        Divide by the lowest of these numbers which is 0.5:
        Methane = 0.5 ÷ 0.5 = 1 mol
        Oxygen = 1 ÷ 0.5 = 2 mol
        Carbon dioxide = 0.5 ÷ 0.5 = 1 mol
        Water = 1 ÷ 0.5 = 2 mol *[1 mark]*
        $CH_4 + 2O_2 \rightarrow CO_2 + 2H_2O$ **[1 mark]**
3.2   Moles of oxygen = 48 g ÷ 32 = 1.5 mol
        Molar ratio of oxygen : carbon dioxide = 2:1
        Moles of carbon dioxide = 1.5 mol ÷ 2 = **0.75 mol** *[3 marks for correct answer, otherwise 1 mark for 1.5 mol of oxygen and 1 mark for molar ratio of 2:1]*
3.3   Molar ratio of CH_4 : H_2O = 1:2
        4 mol of methane will produce **8 mol** of water *[1 mark]*.
3.4   Mass of water = 18 × 8 = **144 g** *[1 mark]*

If you got the equation wrong in 3.1 but used all the right working in parts 3.2, 3.3 and 3.4, you still get the marks, even if you got a different answer to the one here.

Page 41 — Limiting Reactants

1.1   To make sure that all the hydrochloric acid was used up in the reaction *[1 mark]*.
1.2   The limiting reactant is completely used up during a reaction *[1 mark]* and so its quantity limits the amount of product that can be formed *[1 mark]*.
2.1   Molar ratio of copper oxide : copper sulfate = 1:1
        Therefore, 0.50 mol of copper sulfate is produced.
        M_r of copper sulfate = 63.5 + 32 + (16 × 4) = 159.5
        Mass of copper sulfate = 0.50 × 159.5 = **80 g** *[3 marks for correct answer, otherwise 1 mark for 0.50 moles of copper sulfate and 1 mark for M_r of 159.5]*
2.2   The amount of product formed is directly proportional to the amount of limiting reactant *[1 mark]*. So doubling the quantity of the sulfuric acid will double the yield of the copper sulfate *[1 mark]*.
2.3   If only 0.4 mol of copper oxide is present, there will not be enough molecules to react with all the sulfuric acid *[1 mark]*. The copper oxide will be the limiting reactant *[1 mark]* and only 0.4 mol of product will be formed *[1 mark]*.

Pages 42-43 — Gases and Solutions

1.1   Conc. of calcium chloride = 28 g ÷ 0.4 dm^3 = **70 g/dm³**
        [1 mark for correct answer and 1 mark for correct units]
1.2   The concentration of a solution is the amount of a substance in a given volume of a solution *[1 mark]*.
2.1   Mass of CO_2 = (36 ÷ 24) × 44 = **66 g**
        [2 marks for correct answer, otherwise 1 mark for using the correct equation to calculate mass]
2.2   1 mole of any gas at room temperature and pressure has a volume of 24 dm^3 *[1 mark]*. So, 1 mole of CO_2 and 1 mole of O_2 will have the same volume *[1 mark]*.
3.1 M_r of O_2 = 2 × 16 = 32
        Volume of oxygen = (16 g ÷ 32) × 24 = **12 dm³**
        [3 marks for correct answer, otherwise 1 mark for M_r of oxygen, 1 mark for using the correct equation to calculate volume]
3.2   Molar ratio of oxygen : carbon dioxide = 2:1
        Volume of carbon dioxide = 12 ÷ 2 = **6 dm³** *[2 marks for correct answer, otherwise 1 mark for 2:1 molar ratio]*
4.1   Volume of oxygen = 48 ÷ 2 = **24 dm³** *[1 mark]*

As the molar ratio of carbon monoxide to oxygen is 2:1, there must be half the volume of oxygen as there is carbon monoxide.

4.2   Volume of carbon monoxide: (28 ÷ 28) × 24 = 24 dm^3
        Carbon monoxide to oxygen molar ratio = 2:1 so volume of oxygen = 24 ÷ 2 = **12 dm³** *[4 marks for correct answer, otherwise 1 mark for using the formula volume = mass ÷ M_r × 24, 1 mark for correct volume of carbon monoxide and 1 mark for 2:1 molar ratio]*
4.3   24 dm^3 *[1 mark]*

The molar ratio of carbon monoxide to carbon dioxide is 2:2, therefore there is the same volume of carbon dioxide as there is carbon monoxide.

5.1   Moles of Na_2CO_3 = 0.50 mol/dm³ × 0.50 dm³ = **0.25 mol** *[1 mark]*
5.2 M_r of Na_2CO_3 = (23 × 2) + 12 + (16 × 3) = 106
        Mass of 0.25 mol of Na_2CO_3 = 0.25 × 106 = **26.5 g** *[2 marks for correct answer, otherwise 1 mark for correct working]*

Pages 44-45 — Concentration Calculations

1.1 M_r of HCl = 1 + 35.5 = 36.5
        Concentration of HCl = 18.25 ÷ 36.5 = **0.500 mol/dm³**
        [3 marks for correct answer, otherwise 1 mark for M_r of HCl, 1 mark for correct equation to convert the units of concentration]
1.2   Volume of HCl = 25.0 cm^3 ÷ 1000 = 0.0250 dm^3
        Moles of HCl = 0.500 mol/dm³ × 0.0250 dm³ = **0.0125 mol**
        [3 marks for correct answer, otherwise 1 mark for concentration of HCl in dm³ and 1 mark for correct equation to calculate moles]
1.3   Moles of NaOH = **0.0125 mol** *[1 mark]*

The molar ratio of hydrochloric acid to sodium hydroxide is 1:1, so there must be the same number of moles of sodium hydroxide as there are hydrochloric acid.

1.4   Volume of NaOH = 50.0 cm^3 ÷ 1000 = 0.0500 dm^3
        Concentration of NaOH = 0.0125 mol ÷ 0.0500 dm³ = **0.250 mol/dm³**
        [3 marks for correct answer, otherwise 1 mark for volume of NaOH in dm³ and 1 mark for correct equation to calculate concentration]
1.5 M_r of NaOH = 23 + 16 + 1 = 40
        Concentration of NaOH = 0.250 mol × 40 = **10 g/dm³**
        [3 marks for correct answer, otherwise 1 mark for M_r of NaOH and 1 mark for correct equation to convert the units of concentration]
2.1 $Na_2CO_3 + 2HCl \rightarrow 2NaCl + H_2O + CO_2$
        [1 mark for correct symbols, 1 mark for correct balancing]
2.2   Mean volume of HCl = (12.50 + 12.55 + 12.45) ÷ 3 = **12.50 cm³**
        [2 marks for correct answer, otherwise 1 mark for correct equation to calculate the mean]
2.3   Volume of HCl = 12.50 cm^3 ÷ 1000 = 0.01250 dm^3
        Volume of Na_2CO_3 = 25.0 cm^3 ÷ 1000 = 0.0250 dm^3
        Moles of HCl = 0.0125 dm³ × 1.00 mol/dm³ = 0.0125 mol
        Molar ratio of HCl : Na_2CO_3 = 2:1
        Moles of Na_2CO_3 = 0.01250 mol ÷ 2 = 0.006250 mol
        Concentration of Na_2CO_3 = 0.006250 mol ÷ 0.0250 dm³ = **0.250 mol/dm³**
        [6 marks for correct answer, otherwise 1 mark for volumes of HCl and Na_2CO_3 in dm³, 1 mark for correct equation to calculate moles, 1 mark for moles of HCl, 1 mark for moles of Na_2CO_3 and 1 mark for correct equation to calculate concentration]

If answer to question 2.2 is incorrect, but your working is correct here, you still get all the marks, even if you got a different answer.

Page 46 — Atom Economy

1.1 Any two from: e.g. less waste / more sustainable / more profitable *[2 marks — 1 mark for each correct answer]*.

1.2 M_r of ethanol = $(12 \times 2) + (1 \times 6) + 16 = $ **46** *[1 mark]*

1.3 M_r of ethene = $(12 \times 2) + (1 \times 4) = $ **28** *[1 mark]*

1.4 Atom economy = $(28 \div 46) \times 100 = $ **61%**
[2 marks for correct answer, otherwise 1 mark for correct equation for atom economy]

2 Atom economy of reaction using magnesium:
M_r of reactants = $A_r(Mg) + (2 \times M_r(HCl))$
= $24 + (2 \times 36.5) = 97$
Atom economy = $(2 \div 97) \times 100 = $ **2%** *[2 marks for correct answer, otherwise 1 mark for M_r of reactants]*
Atom economy of reaction using zinc:
M_r of reactants = $A_r(Zn) + (2 \times M_r(HCl))$
= $65 + (2 \times 36.5) = 138$
Atom economy = $(2 \div 138) \times 100 = $ **1%** *[2 marks for correct answer, otherwise 1 mark for M_r of reactants]*
More economical reaction: the magnesium reaction *[1 mark]*

Page 47 — Percentage Yield

1.1 Percentage yield = $(1.8 \text{ g} \div 2.4 \text{ g}) \times 100 = $ **75%**
[2 marks for correct answer, otherwise 1 mark for correct method]

1.2 Any one from: e.g. some of the magnesium may not yet have reacted / some product may have been left behind in the crucible *[1 mark]*.

2.1 Mass of N_2 in g = $14 \times 1000 = 14\,000$ g
Number of moles of N_2 = $14\,000 \div (2 \times 14) = 500$ mol
500 mol of N_2 react to produce $2 \times 500 = 1000$ mol of NH_3.
M_r of NH_3 = $14 + (3 \times 1) = 17$
Theoretical yield = moles $\times M_r$ = $1000 \times 17 = 17\,000$ g
= **17 kg**
[4 marks for correct answer, otherwise 1 mark for correct number of moles of N_2, 1 mark for correct number of moles of NH_3 and 1 mark for M_r of NH_3]

2.2 Percentage yield = $(4.5 \text{ kg} \div 17 \text{ kg}) \times 100 = $ **26%**
[2 marks for correct answer, otherwise 1 mark for correct method]

2.3 Any two from: e.g. the reaction is reversible so may not have gone to completion / products may have been lost during the reaction / there may have been side reactions *[1 mark for each correct answer]*.

2.4 Any two from: e.g. to reduce waste / increase sustainability / to reduce cost *[2 marks — 1 mark for each correct answer]*.

Topic 4 — Chemical Changes

Page 48 — Acids and Bases

Warm-up
Universal indicator will turn **red** in strongly acidic solutions and **purple** in strongly alkaline solutions. In a **neutral** solution, Universal indicator will be green. A pH probe attached to a pH meter is **more** accurate than Universal indicator as it displays a numerical value for pH.

1.1 beer *[1 mark]*

1.2 blue / blue-green *[1 mark]*

1.3 H^+ *[1 mark]*

1.4 0 *[1 mark]* – 14 *[1 mark]*

2.1 acid + alkali → salt + water *[1 mark]*

2.2 $H^+_{(aq)} + OH^-_{(aq)} \rightarrow H_2O_{(l)}$ *[1 mark]*
You still get the marks if you didn't include state symbols.

Page 49 — Titrations

Warm-up

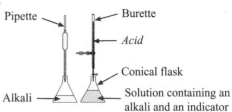

1.1 Universal indicator is the most suitable indicator for use in titrations *[1 mark]*.

1.2 A burette allows an acid/alkali to be added to a solution drop-by-drop *[1 mark]* which helps determine the end-point more accurately *[1 mark]*.

1.3 The titration should be repeated several times to achieve several consistent readings *[1 mark]*. The mean reading should be used to calculate the concentration *[1 mark]*.

Page 50 — Strong Acids and Weak Acids

1.1 A strong acid completely ionises/dissociates in solution *[1 mark]*. A weak acid only partly ionises in solution *[1 mark]*.

1.2 Nitric acid would have a lower pH than ethanoic acid *[1 mark]* because it is a stronger acid/more dissociated/ionised *[1 mark]*, so the concentration of H^+ would be greater *[1 mark]*.
You would also get the marks for using the reverse argument — ethanoic acid would have a higher pH because it is a weaker acid so the concentration of H^+ ions is lower.

1.3 3 *[1 mark]*
As the concentration of H^+ ions in solution decreases by a factor of 10, the pH rises by 1.

1.4 Adding water to the beaker *[1 mark]*.
Adding ethanoic acid to the beaker at the same concentration as the citric acid *[1 mark]*.
Changing the citric acid to carbonic acid of the same concentration *[1 mark]*.

Pages 51-52 — Reactions of Acids

1.1 Neutralisation *[1 mark]*

1.2 Fizzing — Carbon dioxide is produced *[1 mark]*

2.1 sulfuric acid + lithium hydroxide → lithium sulfate + water *[1 mark]*

2.2 $H_2SO_4 + 2LiOH \rightarrow Li_2SO_4 + 2H_2O$ *[1 mark for correct formula of Li_2SO_4, 1 mark for correct balancing]*

2.3 Both reactions produce lithium sulfate and water *[1 mark]*. The reaction between sulfuric acid and lithium carbonate also produces carbon dioxide *[1 mark]*.

3.1 Add zinc oxide to hydrochloric acid until the reaction stops / the excess metal oxide sinks to the bottom *[1 mark]*. Filter the excess solid from the solution using a filter funnel *[1 mark]*. Heat the zinc chloride solution to evaporate some of the water and then leave to cool *[1 mark]*. Filter and dry the crystals that form *[1 mark]*.

3.2 E.g. zinc carbonate *[1 mark]*.
Any other insoluble zinc base or zinc metal also gets a mark.

4   How to grade your answer:

Level 0:   Nothing written worth of credit *[No marks]*.

Level 1:   Some suitable tests are named but it is not clear how the results would enable the solutions to be identified. The chemistry of the tests is not clearly described *[1 to 2 marks]*.

Level 2:   Tests that enable at least one solution to be identified are clearly described, or tests that would enable all solutions to be identified are named but not clearly described *[3 to 4 marks]*.

Level 3:   At least two tests are described together with the expected outcomes. It is clear how these tests would be used to distinguish between all three solutions. The chemistry of the tests is correctly described *[5 to 6 marks]*.

Here are some points your answer may include:

Test the pH of each solution.

The neutral solution/the solution that turns Universal indicator green is the salt.

Add a couple of drops of Universal indicator to the solutions followed by some dilute acid.

The solution containing sodium carbonate will fizz as it reacts with the acid to release carbon dioxide gas as shown by the equation: acid + sodium carbonate → sodium salt + water + carbon dioxide

The solution containing sodium hydroxide will react with acid changing the Universal indicator solution from blue/purple to green, but there won't be any fizzing as no gas is released as shown by the reaction:
acid + sodium hydroxide → sodium salt + water

The solution containing the sodium salt won't react with acid.

Pages 53-54 — The Reactivity Series

1.1   magnesium + hydrochloric acid → magnesium chloride + hydrogen *[1 mark]*

1.2   Positive magnesium ions *[1 mark]*

1.3   It forms positive ions less easily / it's lower down in the reactivity series *[1 mark]*.

1.4   Any one of: e.g. potassium / sodium / lithium / calcium *[1 mark]*.

2.1   metal + water → metal hydroxide + hydrogen *[1 mark]*

2.2 $Ca_{(s)} + 2H_2O_{(l)} \rightarrow Ca(OH)_{2(aq)} + H_{2(g)}$ *[1 mark for each correct product]*

2.3   Any one from: e.g. lithium / sodium / potassium *[1 mark]*
As it is higher in the reactivity series than calcium / loses electrons more easily than calcium / forms positive ions more easily *[1 mark]*.

2.4   potassium, sodium, zinc *[1 mark]*

3.1   When a metal reacts with an acid, the metal forms positive ions *[1 mark]*. The results show that lithium reacts more vigorously with acid than magnesium does *[1 mark]*, so lithium forms positive ions more easily *[1 mark]*.

3.2   A very vigorous fizzing/more vigorous than lithium *[1 mark]*, sodium disappears *[1 mark]*.

3.3   lithium, calcium, copper *[1 mark]*

3.4   It is not possible to tell the difference between magnesium and zinc from these results since both have same reaction with dilute acid *[1 mark]*. E.g. to find which is more reactive, you could find the effect of adding zinc to water *[1 mark]*.

Page 55 — Separating Metals from Metal Oxides

1.1   E.g. gold *[1 mark]*

1.2   Many metals can react with other elements/oxygen to form compounds/oxides *[1 mark]*.

1.3   Reduction is the loss of oxygen *[1 mark]*.

1.4   Magnesium is more reactive than carbon *[1 mark]*.

2.1 $2Fe_2O_3 + 3C \rightarrow 4Fe + 3CO_2$
[1 mark for correct equation, 1 mark for correct balancing]

2.2   Carbon has been oxidised *[1 mark]* as it has gained oxygen during this reaction *[1 mark]*.

2.3   E.g. extracting magnesium would have high energy costs to provide the high temperature and reduced pressure needed *[1 mark]*, but iron extraction doesn't need to be continuously heated *[1 mark]*.

Page 56 — Redox Reactions

1.1   Reduction is the gain of electrons *[1 mark]*.

1.2   zinc chloride + sodium → zinc + sodium chloride *[1 mark]*

1.3   Hydrogen gains electrons *[1 mark]*.

1.4   Chlorine is neither oxidised nor reduced *[1 mark]*.

2.1 $Mg_{(s)} + Fe^{2+}_{(aq)} \rightarrow Mg^{2+}_{(aq)} + Fe_{(s)}$ *[1 mark]*
You still get the marks if you didn't include state symbols.

2.2   No reaction would occur *[1 mark]*. Copper is less reactive than iron so doesn't displace it *[1 mark]*.

Pages 57-59 — Electrolysis

Warm-up

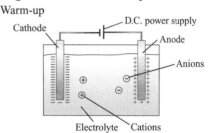

1.1   A liquid or solution that can conduct electricity *[1 mark]*.

1.2   lead bromide → lead + bromine *[1 mark]*

1.3   Lead ions have a positive charge *[1 mark]*. This means they are attracted to the negative cathode *[1 mark]*.

1.4 Br^- *[1 mark]*

1.5   oxidation *[1 mark]*

1.6   So the ions can move to the electrodes *[1 mark]*.

2.1   molten aluminium *[1 mark]*

2.2   To lower the melting point of the electrolyte *[1 mark]*.

2.3   Carbon in the electrodes reacts with oxygen to form carbon dioxide *[1 mark]*, so they degrade over time *[1 mark]*.

3.1   Iron ions, chloride ions, hydrogen ions and hydroxide ions *[1 mark for iron ions and chloride ions, 1 mark for hydrogen ions and hydroxide ions]*.

3.2   At the cathode: hydrogen is discharged.
At the anode: chlorine is discharged *[1 mark]*.

3.3   oxygen *[1 mark]*

3.4   Iron can be extracted via reduction with carbon *[1 mark]*, which is less expensive than electrolysis *[1 mark]*.

4.1 E.g.

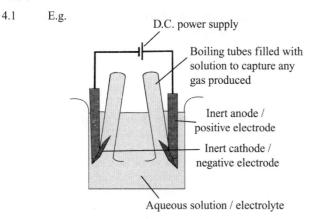

D.C. power supply

Boiling tubes filled with solution to capture any gas produced

Inert anode / positive electrode

Inert cathode / negative electrode

Aqueous solution / electrolyte

[1 mark for power supply, 1 mark for electrodes in solution, 1 mark for boiling tubes over the electrodes, 1 mark for labels]

4.2

Solution	Product at cathode	Product at cathode
$CuCl_2$	Cu	Cl_2
KBr	H_2	Br_2
H_2SO_4	H_2	O_2 and H_2O

[1 mark for each correct answer]

4.3 Potassium is more reactive than hydrogen *[1 mark]* so hydrogen is discharged *[1 mark]*. There are no halide ions *[1 mark]* so oxygen and water are discharged *[1 mark]*.

4.4 Cathode: $2H^+ + 2e^- \rightarrow H_2$ *[1 mark]*
Anode: $4OH^- \rightarrow O_2 + 2H_2O + 4e^-$
/ $4OH^- -- 4e^- \rightarrow O_2 + 2H_2O$
[1 mark]

Topic 5 — Energy Changes

Pages 60-61 — Exothermic and Endothermic Reactions

1 In an endothermic reaction, energy is transferred from the surroundings so the temperature of the surroundings goes down *[1 mark]*.

2.1 endothermic *[1 mark]*

2.2

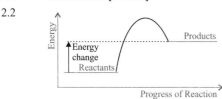

[1 mark for correct curve, 1 mark for energy change]
The curve has to go above the energy of the products and then fall back down. If you didn't do this, you don't get the mark.

2.3 From the surroundings *[1 mark]*.

2.4 It stays the same *[1 mark]*.

2.5 E.g. a sports injury pack *[1 mark]*.

3.1 The activation energy is the minimum amount of energy that reactants must have when they collide with each other in order to react *[1 mark]*. It's shown by the difference between the energy of the reactants and the maximum energy reached by the curve on the reaction profile *[1 mark]*.

3.2 Reaction A is the most suitable reaction *[1 mark]*.
Reaction C is endothermic, so would not give out heat, and couldn't be used to warm your hands *[1 mark]*.
Reaction A has a lower activation energy than Reaction B / gives out more energy than Reaction B *[1 mark]*.

4.1 Any three from: e.g. thermometer / polystyrene cup (and lid) / mass balance / measuring cylinder / beaker filled with cotton wool / stopwatch *[1 mark for each]*.

4.2 How to grade your answer:
Level 0: There is no relevant information *[No marks]*.
Level 1: The method is vague, and misses out important details about how the investigation could be carried out *[1 to 2 marks]*.
Level 2: The method is clear, but misses out a few key details about how the investigation would be carried out or how the variables could be controlled *[3 to 4 marks]*.
Level 3: There is a clear and detailed method that includes ways to reduce energy transfer to the surroundings, and specifies variables that should be controlled throughout the investigation *[5 to 6 marks]*.

Here are some points your answer may include:
Measure out an exact volume of the acid solution into the polystyrene cup.
Record the initial temperature of the acid solution.
Add one metal powder and stir the mixture.
Place a lid on the polystyrene cup to reduce the amount of energy transferred to the surroundings.
Take the temperature of the mixture every 30 seconds and record the highest temperature.
Repeat the experiment for each different metal.
Use the same volume and concentration of acid each time you repeat the experiment.
Make sure the acid starts at the same temperature each time you repeat the experiment.
Use the same number of moles and the same surface area of metal each time you repeat the experiment.

Page 62 — Bond Energies

1.1 Energy to break the bonds = (4 × C–H) + Cl–Cl
= (4 × 413) + 243 = 1652 + 243 = 1895 kJ/mol
Energy produced when bonds form = (3 × C–H) + C–Cl + H–Cl = (3 × 413) + 346 + 432 = 1239 + 346 + 432
= 2017 kJ/mol
Energy change of reaction = Energy to break bonds – Energy produced when bonds form
= 1895 – 2017 = **–122 kJ/mol** *[3 marks for correct answer, otherwise 1 mark for 1895 kJ/mol, 1 mark for 2017 kJ/mol, 1 mark for subtracting energy produced when bonds form from energy needed to break bonds]*

Three of the C–H bonds are unchanged in this reaction. So you could also calculate this by working out just the energy needed to break the C–H and the Cl–Cl bond, and subtracting the energy that's released when the new C–Cl and H–Cl bonds form.

1.2 The reaction is exothermic *[1 mark]* because the energy released when the bonds of the products form is greater than the energy needed to break the bonds of the reactants *[1 mark]*.

2 Total energy needed to break the bonds in the reactants
= H–H + F–F = 436 + 158 = 594 kJ/mol
Energy change of reaction = Energy needed to break bonds – Energy released when bonds form
So, energy released when bonds form = Energy needed to break bonds – Energy change of reaction
= 594 – (–542) = 1136 kJ/mol
Energy released when bonds form = 2 × H–F bond energy
So, H–F bond energy = 1136 ÷ 2 = **568 kJ/mol**
[3 marks for correct answer, otherwise 1 mark for finding the energy needed to break the bonds, 1 mark for finding the energy released by forming bonds.]

Page 63-64 — Cells, Batteries and Fuel Cells

Warm-up

1 False
2 False
3 True

1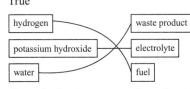

hydrogen — waste product
potassium hydroxide — electrolyte
water — fuel

[2 marks for three correct lines, 1 mark for one correct line]

2.1 Cell B *[1 mark]* as if the two metals are the same then no voltage will be produced / the two metals must be different in order for a voltage to be produced *[1 mark]*.

2.2 The set-up is a battery *[1 mark]*. The voltage will increase to be twice the voltage of Cell B by itself *[1 mark]*.

2.3 The reactants/chemicals in the cells get used up *[1 mark]*, so a voltage is no longer produced *[1 mark]*.

2.4 It can be reversed by connecting the cell to an external electric current *[1 mark]*.

3.1 Reaction equation: $H_2 \rightarrow 2H^+ + 2e^-$ *[1 mark]*
Type of reaction: oxidation *[1 mark]*

3.2 water *[1 mark]*

3.3 Any one of: e.g. hydrogen fuel cells are less polluting to dispose of than rechargeable batteries / there's a limit to how many times a rechargeable battery can be recharged, but this isn't a problem for hydrogen fuel cells / fuel cells store more energy than rechargeable batteries *[1 mark]*.

4.1 C, A, D, B *[2 marks, or 1 mark if 2 correct]*
The greater the voltage of the cell, the more reactive the metal *[1 mark]*. Metal C produces the greatest voltage, so is the most reactive, followed by A, then D, and finally B which produces the lowest voltage *[1 mark]*.

4.2 E.g. the electrolyte *[1 mark]*.

Topic 6 — The Rate and Extent of Chemical Change

Pages 65-67 — Rates of Reaction

1.1 Using a larger volume of the solution, but keeping the concentration the same *[1 mark]*.

1.2 activation energy *[1 mark]*

1.3 A catalyst decreases the activation energy *[1 mark]*.

2 Produced most product: C *[1 mark]*
Finished first: B *[1 mark]*
Started at the slowest rate: A *[1 mark]*

3.1

Volume of hydrogen gas (cm³) vs Time (s)

[1 mark for curve with steeper gradient at the start of the reaction, 1 mark for curve reaching the final volume earlier, 1 mark for final volume being the same as for the other curve]

3.2 The frequency of the collisions *[1 mark]* and the energy of the colliding particles *[1 mark]*.

3.3 There are more particles in a given volume/the particles are closer together *[1 mark]*, so the collisions between particles are more frequent *[1 mark]*.

3.4 The rate would increase *[1 mark]*.

3.5 Smaller pieces have a higher surface area to volume ratio *[1 mark]*. So for the same volume of solid, the particles around it will have more area to work on and collisions will be more frequent *[1 mark]*.

3.6 E.g. changing the temperature / adding a catalyst *[1 mark]*.

4.1 E.g. increasing the volume of the reaction vessel would decrease the pressure of the reacting gases *[1 mark]*. So the particles would be more spread out and would collide less frequently *[1 mark]*, so the reaction rate would decrease *[1 mark]*. Increasing the temperature would cause the particles to move faster, so the frequency of collisions would increase *[1 mark]* and the reaction rate would increase *[1 mark]*.

4.2 It's a catalyst *[1 mark]*.

4.3 The reaction equation won't change *[1 mark]*. Cerium oxide isn't used up in the reaction, so doesn't appear in the reaction equation *[1 mark]*.

4.4

Energy vs Progress of Reaction
Without cerium oxide
With cerium oxide
Reactants
Products

[1 mark for correct relative energies of products and reactants, 1 mark for start and end energies being the same for reactions with and without cerium oxide, 1 mark for reaction with cerium oxide rising to a lower energy than reaction without cerium oxide]

Pages 68-70 — Measuring Rates of Reaction

Warm-up

The rate of a reaction can be measured by dividing the amount of **reactants** used up or the amount of **products** formed by the **time**. To find the rate at a particular time from a graph with a curved line of best fit, you have to find the **gradient** of the **tangent** at that time.

1 mass *[1 mark]*, volume of gas *[1 mark]*

2.1 time taken for the solution to go cloudy *[1 mark]*

2.2 temperature *[1 mark]*

2.3 Any one from: e.g. the concentration of the reactants / the volume of the reactants / the depth of the reaction mixture *[1 mark]*.

2.4 It would be more accurate to measure the volume of gas produced *[1 mark]* as this method less subjective *[1 mark]*.

3.1 E.g. a gas syringe / a measuring cylinder inverted in a bowl of water *[1 mark]*.

3.2

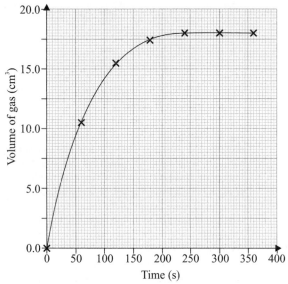

[2 marks for all points plotted correctly, or 1 mark for at least 5 points plotted correctly, 1 mark for line of best fit.]

3.3　　Any value between 210-240 s *[1 mark]*

When no more gas is produced, the reaction has stopped.

3.4　　E.g. Mean rate of reaction = $\dfrac{\text{amount of product formed}}{\text{time for reaction to stop}}$

$= \dfrac{18.0}{240} = \textbf{0.075 cm}^3\textbf{/s}$

[2 marks for correct answer between 0.075-0.086 cm³/s, otherwise 1 mark for correct equation]

If you got the wrong answer in 3.3, but used it correctly here as the change in y, you still get all the marks.

3.5　　E.g. repeat the experiment using the same method *[1 mark]* and check that the results are similar *[1 mark]*.

4.1

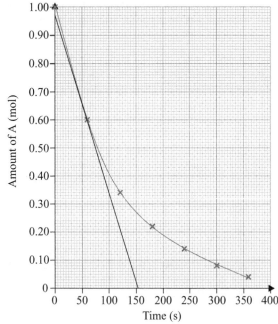

Gradient = $\dfrac{\text{change in } y}{\text{change in } x} = \dfrac{0.97}{153} = \textbf{0.0063 mol/s}$

(allow between 0.0053 mol/s and 0.0073 mol/s)

[4 marks for correct answer, otherwise 1 mark for correctly drawn tangent to curve at 50 s, 1 mark for answer to 2 s.f., 1 mark for correct units]

4.2

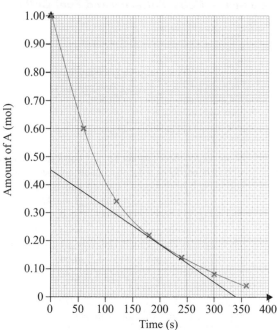

Gradient = $\dfrac{\text{change in } y}{\text{change in } x} = \dfrac{0.45}{340} = \textbf{0.0013 mol/s}$

(allow between 0.0008 mol/s and 0.0018 mol/s)

[4 marks for correct answer, otherwise 1 mark for correctly drawn tangent to curve at 200 s, 1 mark for answer to 2 s.f., 1 mark for correct units]

4.3　　The rate decreases *[1 mark]*. This is because, as the amount of reactant A falls, so does its concentration and so the frequency of collisions between the reactant particles decreases *[1 mark]*.

Page 71 — Reversible Reactions

1.1　　That the reaction is reversible / can go both ways *[1 mark]*.

1.2　　At equilibrium, the rate of the forward reaction is equal to the rate of the backwards reaction *[1 mark]*.

2.1　　It will be exothermic *[1 mark]*. The same amount of energy will be released in the reverse reaction as is taken in by the forward reaction *[1 mark]*.

2.2　　The system has reached equilibrium *[1 mark]*. This mixture contains both blue copper(II) ions and the yellow copper compound, so the colours mix to form green *[1 mark]*.

2.3　　E.g. by changing the temperature / by changing the concentration of one of the reactants *[2 marks — 1 mark for each correct answer]*.

Pages 72-73 — Le Chatelier's Principle

Warm-up

　　　　more reactants
　　　　more reactants
　　　　more products

1.1　　If you change the conditions of a reversible reaction at equilibrium, the system will try to counteract that change *[1 mark]*.

1.2　　E.g. the temperature / the concentration of the reactants *[2 marks — 1 mark for each correct answer]*

2.1　　At higher temperatures there will be more ICl and less ICl₃ / the equilibrium will shift to the left *[1 mark]*. This is because the reverse reaction is endothermic so opposes the increase in temperature *[1 mark]*.

2.2　　There would be more ICl₃ and less ICl *[1 mark]* because the increase in pressure *[1 mark]* causes the equilibrium position to move to the side with the fewest molecules of gas *[1 mark]*.

3.1   At higher temperature there's more product (brown NO_2) in the equilibrium mixture *[1 mark]*. This suggests that the equilibrium has moved to the right/forward direction *[1 mark]*, so the forward reaction is endothermic *[1 mark]*.

From Le Chatelier's principle, you know that increasing the temperature will favour the endothermic reaction as the equilibrium tries to oppose the change. So the forward reaction must be endothermic, as there's more NO_2 in the equilibrium mixture at higher temperatures.

3.2   The mixture would go a darker brown *[1 mark]*, as the decrease in pressure causes the equilibrium to move to the side with the most molecules of gas *[1 mark]*, meaning more NO_2 is formed *[1 mark]*.

4   Observation 1: Increasing amounts of red $FeSCN^{2+}$ are formed, so the solution becomes a darker red *[1 mark]*. When equilibrium is reached, the amount of each substance stops changing, and so does the colour *[1 mark]*.
Observation 2: The concentration of Fe^{3+} initially increases, so the solution becomes more orangey *[1 mark]*. The equilibrium then shifts to make more $FeSCN^{2+}$, so the solution becomes darker red in colour *[1 mark]*.
Observation 3: The concentration of $FeSCN^{2+}$ initially increases, so the solution becomes darker red *[1 mark]*. The equilibrium then shifts to produce more reactants, so the solution becomes more orangey *[1 mark]*.

Topic 7 — Organic Chemistry

Pages 74-75 — Hydrocarbons

Warm-up

Hydrocarbon	Not a hydrocarbon
propane ethene C_2H_6 C_2H_4	butanoic acid CH_3CH_2Cl hydrochloric acid

1.1   A compound that is formed from hydrogen and carbon atoms only *[1 mark]*.
1.2   butane, propane, ethane, methane *[1 mark]*
1.3 C_nH_{2n+2} *[1 mark]*
1.4   hydrocarbon + oxygen → **carbon dioxide** + **water** *[1 mark]*
1.5   oxidised *[1 mark]*
2.1   **B** *[1 mark]*
2.2   **B**, **D**, and **E** *[1 mark]*. They have the general formula C_nH_{2n+2} *[1 mark]*.
2.3   **E** *[1 mark]*. Boiling point increases with increasing molecular size/number of carbons *[1 mark]*.
3.1   Diesel will be more viscous than petrol *[1 mark]*. The higher boiling point of diesel means it contains larger molecules/molecules with longer chains *[1 mark]*.
3.2   Petrol *[1 mark]*. The lower boiling point of petrol means it contains smaller molecules/molecules with shorter chains *[1 mark]*.
3.3 $C_{20}H_{42}$ *[1 mark]*
3.4 $2C_8H_{18} + 25O_2 \rightarrow 16CO_2 + 18H_2O$ *[1 mark for correct formulas of products, 1 mark for balancing]*

Any correct balance of the equation is correct, e.g. $C_8H_{18} + 12\frac{1}{2}O_2 \rightarrow 8CO_2 + 9H_2O$.

Page 76 — Fractional Distillation

1.1   The remains of ancient organisms/plankton *[1 mark]*.
1.2   A resource which is being used quicker than it is being replaced so will run out eventually *[1 mark]*
1.3   alkanes *[1 mark]*
2.1   boiling point *[1 mark]*

2.2   The fractionating column is hot at the bottom and cool at the top *[1 mark]*. So longer hydrocarbons, which have higher boiling points, will condense and be drained off near the bottom *[1 mark]*. Meanwhile, shorter hydrocarbons, with lower boiling points, will condense and be drained off further up the column *[1 mark]*.
2.3   They contain similar numbers of carbon atoms / they have a similar chain length *[1 mark]*.

Pages 77-78 — Uses and Cracking of Crude Oil

1.1   Any two from: e.g. solvents / lubricants / polymers / detergents *[2 marks — 1 mark for each correct answer]*
1.2   cracking *[1 mark]*
1.3   E.g. shorter chain hydrocarbons are more useful/can be used for more applications *[1 mark]*.
2.1   thermal decomposition / endothermic *[1 mark]*
2.2   Hydrocarbons are vaporised / heated to form gases *[1 mark]*. The vapours are then passed over a hot catalyst / the vapours are mixed with steam and heated to very high temperatures *[1 mark]*.
2.3   E.g. $C_{10}H_{22} \rightarrow C_7H_{16} + C_3H_6$ *[1 mark]*

Cracking equations must always be balanced and have a shorter alkane and an alkene on the right-hand side.

3.1 C_7H_{16} *[1 mark]*
3.2

[1 mark for correct number of carbons, 1 mark for correct displayed formula]

3.3   E.g. to produce polymers / as a starting material for other chemicals *[1 mark]*.
4   How to grade your answer:
   Level 0: Nothing written worth of credit *[No marks]*.
   Level 1: Basic outline of how some fractions are processed but lacking detail. Some mention of the uses of cracking products *[1 to 2 marks]*.
   Level 2: Reason for cracking explained and some detail given about the process. The uses of cracking products are covered in detail *[3 to 4 marks]*.
   Level 3: Reasons for cracking and the process of cracking are explained in detail, including an accurate balanced symbol or word equation. Examples given of the uses of the products of cracking *[5 to 6 marks]*.

Here are some points your answer may include:
<u>Reasons for cracking</u>
There is a higher demand for short chain hydrocarbons as these make good fuels.
Long chain hydrocarbons are less useful than short chain hydrocarbons, so there is less demand for them.
Cracking helps the supply of short chain hydrocarbons to meet the demand.
<u>Cracking process</u>
The long chain hydrocarbons are heated and vaporised.
The vapours are passed over a hot catalyst / mixed with steam and heated to a high temperature so that they thermally decompose.
Any relevant word equation: e.g. decane → octane + ethene
Any relevant balanced equation:
e.g. $C_{12}H_{26} \rightarrow C_8H_{18} + 2C_2H_4$
<u>Uses of cracking products</u>
The products of cracking are useful as fuels.
Alkenes are used as a starting material when making lots of other compounds and can be used to make polymers.

Page 79 — Alkenes

1.1   A hydrocarbon with a double carbon-carbon bond *[1 mark]*.

1.2 C_nH_{2n} *[1 mark]*

1.3

$$H_2C=CH-CH_3$$

[1 mark]

2.1   **B** and **E** *[1 mark]*

2.2 C_6H_{12} *[1 mark]*

2.3   **B** undergoes incomplete combustion in low amounts of oxygen *[1 mark]*.
Any balanced incomplete combustion equation:
e.g. $2C_2H_4 + 5O_2 \rightarrow 2CO_2 + 2CO + 4H_2O$ *[1 mark for correct formulas of reactants and products, 1 mark for balancing]*

Pages 80-81 — Reactions of Alkenes

Warm-up

Alkenes generally react via **addition** reactions to form a variety of compounds. Alkenes can react with steam to form **alcohols**. For example, **ethene** can be mixed with steam and passed over a catalyst to form ethanol.

1.1   C=C / carbon-carbon double bond *[1 mark]*

1.2   They have the same functional group *[1 mark]*.

1.3   The double bond opens up to leave a single bond *[1 mark]*. A new atom is added to each of the C=C carbons *[1 mark]*.

2.1 C_3H_6 *[1 mark]*

2.2   a catalyst / nickel *[1 mark]*

2.3   Add bromine water *[1 mark]*. With propene it will change from orange to colourless *[1 mark]*. Propane will not react and there will be no colour change *[1 mark]*.

3.1

$$H_3C-CHBr-CHBr-CH_3$$

[1 mark]

3.2   The product of the reaction is saturated *[1 mark]*. It doesn't contain any carbon-carbon double bonds *[1 mark]*.

4.1   E.g.

$$H_3C-CHCl-CH_2I$$

[1 mark]

You still get the mark if you have the chlorine and iodine atoms swapped around.

4.2

$$H_3C-CH_2-CH_2-OH$$

[1 mark]

and

$$H_3C-CHOH-CH_3$$

[1 mark]

Page 82 — Addition Polymers

1.1   ethene *[1 mark]*

1.2   addition polymerisation *[1 mark]*

1.3   carbon-carbon double bond / C=C *[1 mark]*

2.1

$$\left(CHCl-CH_2 \right)_n$$

[1 mark]

2.2

$$H_2C=CHCl$$

[1 mark]

2.3   chloroethene *[1 mark]*

Pages 83-84 — Alcohols

1.1   –OH *[1 mark]*

1.2   Any two from: e.g. fuels / solvents / in alcoholic drinks *[2 marks — 1 mark for each correct answer]*.

1.3   Methanol dissolves. The indicator remains green *[1 mark]*.

1.4 CO_2 and H_2O *[1 mark]*

2.1   Ethanoic acid *[1 mark]*, a carboxylic acid *[1 mark]*.

2.2   Yeast *[1 mark]*. The reaction conditions are a temperature of approximately 37 °C *[1 mark]*, absence of oxygen / anaerobic conditions *[1 mark]* and a slightly acidic pH *[1 mark]*.

3.1

$$H_3C-O-H$$

[1 mark]

3.2   Methanol only has only one carbon atom where as butanol has four/methanol has the formula CH_3OH whereas butanol has the formula of C_4H_9OH *[1 mark]*. Both methanol and butanol contain an –OH group *[1 mark]*.

3.3   They both contain the same functional (–OH) group *[1 mark]*.

3.4   hydrogen *[1 mark]*

4.1 $C_2H_4(OH)_2/C_2H_6O_2$ *[1 mark]*

4.2   E.g. react it with sodium *[1 mark]*.

4.3 $2C_2H_4(OH)_2 + 5O_2 \rightarrow 4CO_2 + 6H_2O$ *[1 mark for the formulas of the reactants and products, 1 mark for balancing]*
Any correct balance of the equation is correct, e.g. $C_2H_4(OH)_2 + 2\frac{1}{2}O_2 \rightarrow 2CO_2 + 3H_2O$.

4.4   7/neutral *[1 mark]*

Pages 85-86 — Carboxylic Acids

1.1   –COOH *[1 mark]*

1.2   Methanoic acid would dissolve *[1 mark]* and the Universal indicator would change to red/orange *[1 mark]*.

1.3   carbon dioxide/CO_2 *[1 mark]*

2.1   butanoic acid *[1 mark]*, $C_3H_7COOH/C_4H_8O_2$ *[1 mark]*

2.2   It does not ionise completely when dissolved in water *[1 mark]*

2.3   An acid catalyst *[1 mark]*.

2.4   ester *[1 mark]*

3.1   Accept anywhere between 2 and 6 *[1 mark]*.

3.2   magnesium carbonate/$MgCO_3$ *[1 mark]*

3.3   ethanol/C_2H_5OH *[1 mark]*

4.1 Na_2CO_3 *[1 mark]*

4.2 H_2O *[1 mark]*

4.3   **D** *[1 mark]*. Propanoic acid is a weak acid *[1 mark]* and therefore weakly ionised *[1 mark]*.

4.4   Carbon dioxide/CO_2 *[1 mark]* and water/H_2O *[1 mark]*.

Page 87 — Condensation Polymers

1.1   A small molecule is lost when condensation polymers are formed *[1 mark]*.

1.2   two *[1 mark]*

1.3   E.g. carboxylic acids *[1 mark]* and alcohols / amines *[1 mark]*.

2.1 H_2O *[1 mark]*

2.2

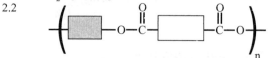

[1 mark for correct ester link connecting monomers, 1 mark for rest of the structure being correct]

2.3   No *[1 mark]*. As two different functional groups that react with each other are needed *[1 mark]*.

Page 88 — Naturally Occurring Polymers

1.1 two *[1 mark]*
1.2 Any one from: carboxylic acid / amine *[1 mark]*
1.3 condensation polymerisation *[1 mark]*
1.4 water *[1 mark]*
2.1 sugars *[1 mark]*
2.2 e.g. cellulose *[1 mark]*
2.3 DNA contains genetic instructions *[1 mark]* for the operation and functioning of living organisms and viruses *[1 mark]*.
2.4 Two polymer chains *[1 mark]* made from four different nucleotide monomers *[1 mark]* linked together by cross links to give a double helix structure *[1 mark]*.

Topic 8 — Chemical Analysis

Page 89 — Purity and Formulations

1.1 A single element or compound not mixed with any other substance *[1 mark]*.
1.2 Sample **A** *[1 mark]*. The purer the substance, the smaller the range of the melting point / purer substances melt at higher temperatures than impure substances *[1 mark]*.
1.3 Sample **A** *[1 mark]*.
2.1 It is a mixture that has been designed to have a precise purpose *[1 mark]*. Each of the components is present in a measured quantity *[1 mark]* and contributes to the properties of the formulation *[1 mark]*.
2.2 By making sure each component in the mixture is always present in exactly the same quantity *[1 mark]*.
2.3 Any one from: e.g. medicines / cleaning products / fuels / cosmetics / fertilisers / metal alloys *[1 mark]*.

Pages 90-91 — Paper Chromatography

Warm-up

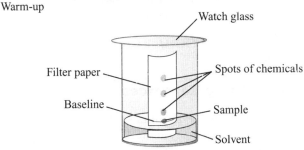

1.1 **E**: $R_f = \dfrac{\text{distance travelled by substance}}{\text{distance travelled by solvent}} = \dfrac{3.6}{9.5} = \mathbf{0.38}$
[2 marks for correct answer, otherwise 1 mark for using correct equation to calculate R_f]
F: $R_f = \dfrac{\text{distance travelled by substance}}{\text{distance travelled by solvent}} = \dfrac{8.0}{9.5} = \mathbf{0.84}$
[2 marks for correct answer, otherwise 1 mark for using correct equation to calculate R_f]
1.2 E.g. they're distributed differently between the mobile phase and the stationary phase *[1 mark]*.
1.3 They're all pure substances *[1 mark]*.
1.4 **D** and **E** *[1 mark]*.
2.1 E.g. to stop any solvent evaporating *[1 mark]*.
2.2 **A** spends more time in the mobile phase compared to the stationary phase than **B** does *[1 mark]*.
2.3 **B** and **C** *[1 mark]*.
2.4 The student is incorrect *[1 mark]*. Substances have different R_f values in different solvents as the attraction between the substance and solvent changes *[1 mark]*.
2.5 It suggest that there are at least 3 substances in **W** *[1 mark]*.
2.6 There were only two spots in the chromatogram shown because two of the substances in **W** are similarly distributed between the mobile phase/water and stationary phase / they had similar R_f values *[1 mark]*.

Pages 92-93 — Tests for Gases and Anions

1.1 A burning splint which results in a popping noise *[1 mark]*.
1.2 oxygen/O_2 *[1 mark]*
1.3 Bubble the gas through an aqueous solution of calcium hydroxide/limewater *[1 mark]*. The limewater turns milky/cloudy *[1 mark]*.
1.4 Damp litmus paper *[1 mark]* put into the gas. If chlorine gas is present, the paper is bleached and turns white *[1 mark]*.
2.1 Potassium, K^+ *[1 mark]*
2.2 Clean a platinum wire loop *[1 mark]* by dipping it in some dilute HCl and then placing it in a blue flame from a Bunsen burner until it burns without colour *[1 mark]*. Dip the loop into the sample you want to test and put it back into the flame *[1 mark]*. Record the colour of the flame *[1 mark]*.
2.3 yellow *[1 mark]*
3.1 sulfate/SO_4^{2-} *[1 mark]*
3.2 $Ba^{2+}_{(aq)} + SO_4^{2-}_{(aq)} \rightarrow BaSO_{4(s)}$ *[1 mark for balanced equation, 1 mark for state symbols]*
3.3 copper(II)/Cu^{2+} *[1 mark]*
3.4 copper sulfate/$CuSO_4$ *[1 mark]*
4.1 Substance **P**: CaI_2/calcium iodide *[1 mark for correct anion, 1 mark for correct cation]*
Substance **R**: LiBr/lithium bromide *[1 mark for correct anion, 1 mark for correct cation]*
4.2 The anion cannot be identified *[1 mark]*.
4.3 **D**: white precipitate *[1 mark]*
E: green *[1 mark]*
4.4 The flame colours of some ions may be hidden by / mixed with the colour of others *[1 mark]*.

Page 94 — Flame Emission Spectroscopy

1.1 Any two from: e.g. faster / more sensitive / more accurate *[2 marks —1 mark for each correct answer]*
1.2 The identity *[1 mark]* and the concentration of metal ions in solution *[1 mark]*.
1.3 line spectra *[1 mark]*
2.1 A sample is placed in a flame and as the ions in the sample heat up they transfer energy as light *[1 mark]*. This light passes through a spectroscope and produces a line spectrum specific to that ion *[1 mark]*.
2.2 Metal **A** and metal **C** *[1 mark]*.

Topic 9 — Chemistry of the Atmosphere

Pages 95-96 — The Evolution of the Atmosphere

Warm-up

1 False
2 True
3 True
4 False

1.1 One-fifth oxygen and four-fifths nitrogen *[1 mark]*.
1.2 Any two from: e.g. carbon dioxide / water vapour / named noble gas *[2 marks — 1 mark for each correct answer]*
1.3 By algae and plants photosynthesising *[1 mark]*.
1.4 By volcanic activity *[1 mark]*.
1.5 200 million years *[1 mark]*
2.1 E.g. photosynthesis by plants and algae / carbon dioxide dissolved in the oceans *[1 mark]*.
2.2 From matter that is buried and compressed over millions of years *[1 mark]*.
2.3 Coal: from thick plant deposits *[1 mark]*.
Limestone: from calcium carbonate deposits from the shells and skeletons of marine organisms *[1 mark]*.
3.1 E.g. the long timescale means there's a lack of evidence *[1 mark]*.
3.2 $6CO_2 + 6H_2O \rightarrow C_6H_{12}O_6 + 6O_2$ *[1 mark]*

3.3   Oxygen is produced by photosynthesis *[1 mark]* and there are no plants or algae / there isn't any photosynthesis *[1 mark]* on Mars.

3.4   The fact that the red beds formed about 2 billion years ago suggests that before this time there wasn't enough oxygen in the air for iron oxide to form / from this time there was enough oxygen in the air for iron oxide to form *[1 mark]*.

Pages 97-98 — Greenhouse Gases and Climate Change

1.1   Nitrogen *[1 mark]*

1.2   They help to keep Earth warm *[1 mark]*.

1.3   Any two from: e.g. deforestation / burning fossil fuels / agriculture / producing waste *[2 marks — 1 mark for each correct answer]*

2.1   Greenhouse gases absorb long-wave (thermal) radiation *[1 mark]* reflected from Earth's surface *[1 mark]*. They then reradiate this thermal radiation in all directions, including back towards Earth, helping to warm the atmosphere *[1 mark]*.

2.2   E.g. flooding *[1 mark]* and coastal erosion *[1 mark]*.

2.3   Any one from: e.g. changes in rainfall patterns / the ability of certain regions to produce food might be affected / the frequency/severity of storms might increase / the distribution of wild species might change *[1 mark]*.

3   How to grade your answer:
Level 0:   There is no relevant information *[No marks]*.
Level 1:   Unstructured and no logic. The trends in the variables are described but reasons are not given *[1 to 2 marks]*.
Level 2:   Some structure and logic but lacking clarity. The trends in the variables are described and there is some explanation of how the increase in carbon dioxide may have come about and how this might be linked to temperature *[3 to 4 marks]*.
Level 3:   Clear, logical answer. The trends in the variables are described and there is a clear explanation of how the increase in carbon dioxide may have come about and how this may be linked to temperature *[5 to 6 marks]*.

Here are some points your answer may include:
The graph shows an increase in carbon dioxide levels in the atmosphere between 1960 and 2015.
The increase in carbon dioxide levels is likely to be due to human activities which release carbon dioxide into the atmosphere.
These activities include increased burning of fossil fuels, increased deforestation and increased waste production.
The graph shows that the increase in carbon dioxide appears to correlate with an increase in global temperatures.
The increase in global temperatures is likely to be due to the increase in carbon dioxide in the atmosphere, as carbon dioxide is a greenhouse gas so helps to keep Earth warm.

4.1   The global warming potential for methane is significantly greater than for carbon dioxide *[1 mark]*.

4.2   It has a very high global warming potential compared to other gases *[1 mark]* and stays in the atmosphere for a long time *[1 mark]*.

Page 99 — Carbon Footprints

1.1   A measure of the amount of carbon dioxide and other greenhouse gases *[1 mark]* released over the full life cycle of something *[1 mark]*.

1.2   E.g. using renewable or nuclear energy sources *[1 mark]* and using more energy efficient appliances *[1 mark]*.

1.3   E.g. lack of education / reluctance to change their lifestyle / cost of changing lifestyle *[1 mark]*.

2.1   Any two from, e.g: specialist equipment is needed to capture the carbon dioxide / it's expensive to capture and store the carbon dioxide / it could be difficult to find suitable places to store the carbon dioxide *[2 marks — 1 mark for each correct answer]*.

2.2   E.g. governments could tax companies based on the amount of greenhouse gases they emit *[1 mark]*. They could also put a cap on the emissions produced by a company *[1 mark]*. Governments might be reluctant to impose these methods if they think it will affect economic growth / could impact on people's well-being *[1 mark]*, especially if other countries aren't using these methods either / the country is still developing *[1 mark]*.

Page 100 — Air Pollution

1.1   Coal can contain sulfur impurities *[1 mark]*

1.2   Acid rain: sulfur dioxide / nitrogen oxides/nitrogen monoxide/nitrogen dioxide/dinitrogen monoxide *[1 mark]* Global dimming: e.g. (carbon) particulates *[1 mark]*

1.3   Any two from: e.g. damage to plants / buildings / statues / corrodes metals *[2 marks — 1 mark for each correct answer]*.

2.1   The reaction of nitrogen and oxygen from the air *[1 mark]* at the high temperatures produced by combustion *[1 mark]*.

2.2   Nitrogen oxides cause respiratory problems *[1 mark]* and contribute to acid rain *[1 mark]*.

2.3   E.g. they can cause respiratory problems *[1 mark]*.

2.4   Carbon monoxide *[1 mark]*. It is colourless and odourless *[1 mark]*.

Topic 10 — Using Resources

Page 101 — Ceramics, Polymers and Composites

1.1   limestone, sand, sodium carbonate *[3 marks — 1 mark for each correct answer]*

1.2   Because borosilicate glass has a higher melting point than soda-lime glass *[1 mark]*.

1.3   Wet clay is shaped *[1 mark]* then fired at a high temperature *[1 mark]*.

2.1   They are made at different temperatures/pressures *[1 mark]* with a different catalyst *[1 mark]*.

2.2   thermosoftening *[1 mark]*

2.3   Poly(ethene) chains are entwined together with weak forces between the chains *[1 mark]*. Polyester resin can form crosslinks between polymer chains *[1 mark]*.

2.4   Composites consist of fibres/fragments of a material known as the reinforcement *[1 mark]* surrounded by a matrix/binder *[1 mark]*.

Page 102 — Properties of Materials

1.1   Bronze — Copper and tin *[1 mark]*
Steel — Iron and carbon *[1 mark]*
Brass — Copper and zinc *[1 mark]*

1.2   Any one from: e.g. water taps / door fittings *[1 mark]*

2.1   strength increases *[1 mark]*

2.2   high carbon steel *[1 mark]*

2.3   Aluminium is much less dense than the other metals *[1 mark]*.

Page 103 — Corrosion

1.1 iron + water + oxygen → hydrated iron(III) oxide *[1 mark]*

1.2 Galvanising is coating iron with zinc *[1 mark]*. Zinc is more reactive than iron *[1 mark]* so, unlike other methods of barrier protection, even if the coating is scratched it will still prevent the iron from rusting as it will oxidise before iron does *[1 mark]*.

1.3 Any three from: e.g. painting / coating with grease / electroplating / using a barrier / sacrificial protection *[3 marks — 1 mark for each correct answer]*

2.1 The reaction of a material with substances in its environment so it is gradually destroyed *[1 mark]*.

2.2 The oxide that forms when aluminium reacts in the air forms a protective layer over the surface of the metal *[1 mark]*, preventing chemicals reaching the rest of the metal and reacting further *[1 mark]*.

2.3 Oxygen and water react with the magnesium instead of the steel *[1 mark]*.

Page 104 — Finite and Renewable Resources

1.1 Coal *[1 mark]*. It does not form fast enough to be considered replaceable *[1 mark]*.

1.2 A resource that reforms at a similar rate to, or faster, than humans can use it *[1 mark]*.

2.1 E.g. the development of fertilisers has meant higher yields of crops *[1 mark]*.

2.2 Any one from: e.g. synthetic rubber has replaced natural rubber / poly(ester) has replaced cotton in clothes / bricks are used instead of timber in construction *[1 mark]*.

3 Any one advantage from: e.g. allows useful products to be made / provides jobs / brings money into the area *[1 mark]*. Any one disadvantage: e.g. uses large amounts of energy / scars the landscape / produces lots of waste / destroys habitats *[1 mark]*.

Pages 105-106 — Reuse and Recycling

1.1 An approach to development that takes account of the needs of present society *[1 mark]* while not damaging the lives of future generations *[1 mark]*.

1.2 E.g. chemists can develop and adapt processes that use less resources/do less damage to the environment *[1 mark]*. For example, chemists have developed catalysts that reduce the amount of energy required for industrial processes *[1 mark]*.

2.1 The raw materials for the jute bag are more sustainable *[1 mark]* as plant fibres are a renewable resource, whilst crude oil is a finite resource *[1 mark]*.

2.2 The production of the poly(ethene) bag is more sustainable *[1 mark]* as it needs less energy to be produced from its raw materials than the jute bag *[1 mark]*.

2.3 The jute bag can be reused and the poly(ethene) bag can be recycled, improving both their sustainability *[1 mark]*. However, the jute bag is more sustainable if the bags are disposed of in landfill *[1 mark]*, as it is biodegradeable, whilst the poly(ethene) bag isn't *[1 mark]*.

3.1 Any two from: e.g. often uses less energy / conserves the amount of raw materials on Earth / cuts down on waste sent to landfill *[2 marks — 1 mark for each correct answer]*.

3.2 Any one from: e.g. glass / metal *[1 mark]*
E.g. glass is crushed and melted down to form other glass products/other purpose / metal is melted and cast into the shape of a new product *[1 mark]*.

3.3 reusing *[1 mark]*

4.1 Plants are grown on soil containing copper compounds *[1 mark]*, so as they grow, copper builds up in their leaves *[1 mark]*. The plants are burned *[1 mark]*. The resulting ash contains the copper compounds *[1 mark]*.

4.2 By electrolysis of a solution containing the copper compounds *[1 mark]* or by displacement using scrap iron *[1 mark]*.

4.3 Copper is a finite resource *[1 mark]* and will eventually run out *[1 mark]*. Recycling copper makes it more sustainable *[1 mark]*.

Page 107 — Life Cycle Assessments

Warm-up

Getting the Raw Materials — Coal being mined from the ground.
Manufacturing and Packaging — Books being made from wood pulp.
Using the Product — A car using fuel while driving.
Product Disposal — Plastic bags going on to landfill.

1.1 Any two from: e.g. if a product is disposed of in landfill sites, it will take up space / may pollute land/water / energy is used to transport waste to landfill / pollution can be caused by incineration *[2 marks — 1 mark for each correct answer]*.

1.2 Any one from: e.g. energy / water / some natural resources / certain types of waste *[1 mark]*

1.3 They can be subjective / they are difficult to measure *[1 mark]*.

1.4 No *[1 mark]*. Some elements of the LCA are not objective/ require the assessors to make value judgements/cannot be quantified reliably *[1 mark]*, therefore different people are likely to make a different judgement/estimate *[1 mark]*.

1.5 Selective LCAs could be written so they only show elements that support a company's claims / they could be biased *[1 mark]* in order to give them positive advertising *[1 mark]*.

Pages 108-109 — Potable Water

Warm-up

1 False
2 True
3 False

1.1 pure water *[1 mark]*

1.2 e.g. from the ground / lakes / rivers *[1 mark]*.

1.3 passing water through filter beds —- solid waste *[1 mark]* sterilisation — microbes *[1 mark]*

1.4 E.g. chlorine, ozone, ultraviolet light *[3 marks — 1 mark for each correct answer]*.

2.1 **A**: Bunsen burner *[1 mark]*
B: round bottom flask *[1 mark]*
C: thermometer *[1 mark]*
D: condenser *[1 mark]*

2.2 Pour the salt water into the flask and secure it on top of a tripod *[1 mark]*. Connect the condenser to a supply of cold water *[1 mark]* that goes in at the bottom and out at the top *[1 mark]*. Heat the flask and allow the water to boil *[1 mark]*. Collect the water running out of the condenser in a beaker *[1 mark]*.

2.3 Reverse osmosis / a method which uses membranes *[1 mark]*

2.4 Desalination requires a lot of energy compared to the filtration and sterilisation of fresh water *[1 mark]*. Since the UK has a plentiful supply of fresh water there is no need to use desalination processes *[1 mark]*.

Page 110 — Waste Water Treatment

1.1 organic matter, harmful microbes *[2 marks — 1 mark for each correct answer]*

1.2 It may contain harmful chemicals which need to be removed *[1 mark]*.

2.1 To remove grit *[1 mark]* and large bits of material/twigs/ plastic bags *[1 mark]*.

2.2 Substance **A**: sludge *[1 mark]*
Substance **B**: effluent *[1 mark]*

2.3 anaerobic digestion *[1 mark]*

Page 111 — The Haber Process

1.1 Nitrogen, Hydrogen *[1 mark for both]*

1.2 ammonia *[1 mark]*

1.3 Because ammonia is used to make fertilisers *[1 mark]*.

2.1 Low temperature *[1 mark]*. Low temperatures cause the position of equilibrium to shift in favour of the exothermic, forward reaction *[1 mark]* which means more ammonia is produced/you get a higher yield of ammonia *[1 mark]*.

2.2 A higher temperature is used to get a reasonable rate of reaction *[1 mark]*.

2.3 High pressure causes the yield to increase *[1 mark]*. There are fewer moles of gas in the products than reactants/ on the right hand side of the equation than on the left hand side *[1 mark]*. Since high pressure favours the production of fewer moles of gas, the position of equilibrium would move right/to the product side as pressure is increased *[1 mark]*.

2.4 Any one from: e.g. safety / cost / rate *[1 mark for any correct answer]*.

Page 112 — NPK Fertilisers

1.1 potassium chloride, potassium sulfate *[1 mark for each correct answer]*

1.2 They are mined *[1 mark]*.

2.1 calcium nitrate *[1 mark]*

2.2 Phosphate rock + sulfuric acid: calcium phosphate *[1 mark]* and calcium sulfate *[1 mark]*.
Phosphate rock + phosphoric acid: calcium phosphate *[1 mark]*.

2.3 The reaction is carried out at lower concentrations in the lab so that it's safer for the person carrying it out *[1 mark]*. Crystallisation isn't used in industry as it's very slow *[1 mark]*.

Mixed Questions

Pages 113-124 — Mixed Questions

1.1

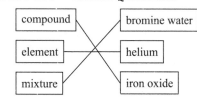

[2 marks if all three correct, otherwise 1 mark if 1 correct]

1.2 Mixtures with a precise purpose *[1 mark]* that are made by following a formula / a recipe *[1 mark]*.

2.1 Dissolve the rock salt in water and filter *[1 mark]*.

2.2 It contains two elements/more than one element in fixed proportions *[1 mark]* held together by chemical bonds *[1 mark]*.

2.3 ionic *[1 mark]*

3.1 Group: 6 *[1 mark]*
Explanation: There are 6 electrons in the outer shell *[1 mark]*.

3.2 2– ions *[1 mark]*, as oxygen atoms need to gain two electrons to get a full outer shell *[1 mark]*.

3.3 Oxidation *[1 mark]*

4.1

[1 mark for shared pair of electrons, 1 mark for six further electrons in the outer shell of each chlorine atom]

4.2 E.g. atoms with the same number of protons / of the same element / with the same atomic number *[1 mark]* with different numbers of neutrons / different mass numbers *[1 mark]*.

4.3 Hold a piece of damp litmus paper in the unknown gas *[1 mark]*. It will be bleached white in the presence of chlorine *[1 mark]*.

4.4 Chlorine is more reactive than iodine *[1 mark]*, so would displace iodine from sodium iodide solution / the solution would go from colourless to brown *[1 mark]*.

5.1 endothermic *[1 mark]*

5.2 higher *[1 mark]*

5.3 It takes more energy to break the bonds in the reactants than is released when the bonds in the products form *[1 mark]*, so overall energy is taken in from the surroundings *[1 mark]*.

5.4 E.g. in a sports injury pack *[1 mark]*.

6.1 alkanes *[1 mark]*

6.2 (fractional) distillation *[1 mark]*

6.3 cracking *[1 mark]*

6.4 Decane *[1 mark]*, because the molecules are bigger *[1 mark]*, so will have stronger intermolecular forces / more energy is needed to break the forces between the molecules *[1 mark]*.

6.5 $C_5H_{12} + 8O_2 \rightarrow 5CO_2 + 6H_2O$ *[1 mark for correct reactants and products, 1 mark for balancing]*

7.1 The electrons in the outer shell *[1 mark]* of the metal atoms are delocalised *[1 mark]*. There is strong electrostatic attraction between the positive metal ions and the shared negative electrons *[1 mark]*.

7.2 Iron: solid *[1 mark]*. Silver: liquid *[1 mark]*

7.3 Iron *[1 mark]*, because it has a higher melting/boiling point *[1 mark]*, so more energy is needed to break the bonds *[1 mark]*.

8.1 The reactant that is used up first / limits the amount of product that's formed *[1 mark]*.

8.2 $M_r(LiOH) = A_r(Li) + A_r(O) + A_r(H) = 7 + 16 + 1 = \textbf{24}$ *[1 mark]*

8.3 Number of moles = mass ÷ molar mass = 1.75 ÷ 7 = **0.25 mol** *[2 marks for correct answer, otherwise 1 mark for using the correct equation to calculate moles]*

8.4 From the reaction equation, 0.50 mol Li forms 0.50 mol LiOH.
Mass of LiOH = number of moles × molar mass = 0.50 × 24 = **12 g** *[3 marks for correct answer, otherwise 1 mark for number of moles of LiOH produced, 1 mark for using the correct equation to calculate mass]*

9.1 It described atoms as having a tiny, positively charged nucleus at the centre *[1 mark]*, surrounded by a cloud of electrons *[1 mark]*.

9.2 Atoms consist of a small nucleus *[1 mark]* which contains the protons and neutrons *[1 mark]*. The electrons orbit the nucleus in fixed energy levels/shells *[1 mark]*.

10.1 The particles in a gas expand to fill any container they're in *[1 mark]*. So the particles of carbon dioxide formed will expand out of the unsealed reaction vessel *[1 mark]*, causing the mass of substance inside the reaction vessel to decrease *[1 mark]*.

10.2 E.g. add a set volume and concentration of hydrochloric acid to the reaction vessel *[1 mark]*. Add a set volume and concentration of sodium carbonate solution *[1 mark]*, connect the reaction flask to a gas syringe *[1 mark]* and start the stop-watch *[1 mark]*. Record the volume of gas collected at regular intervals until the reaction is finished *[1 mark]*. Repeat the experiment, keeping everything the same except for the concentration of acid *[1 mark]*.

10.3 Change in volume = 12.0 cm³
Mean rate of reaction = $\dfrac{\text{amount of product formed}}{\text{time}} = \dfrac{12.0}{30}$ = **0.40 cm³/s** *[2 marks for correct answer, otherwise 1 mark for using the correct equation to calculate rate]*

11.1 Any two from: e.g. it dissolved in oceans / photosynthesis / trapped in rocks and fossil fuels *[2 marks — 1 mark for each correct answer]*

11.2 E.g. methane *[1 mark]*. It is increasing due to more agriculture / waste production *[1 mark]*.

11.3 How to grade your answer:

Level 0: There is no relevant information. *[No marks]*

Level 1: There are a few examples of other pollutant gases, but little discussion of how they are made or what their impacts could be. *[1 to 2 marks]*

Level 2: There are a number of examples of other pollutant gases, with some discussion of how they are made and what their impacts could be. *[3 to 4 marks]*

Level 3: There are a number of examples of other pollutant gases, with a detailed discussion of how they are made and what their impacts could be. *[5 to 6 marks]*

Here are some points your answer may include:

Other pollutant gases include carbon monoxide, sulfur dioxide and nitrogen oxides.

Carbon monoxide is produced when fuels undergo incomplete combustion.

Carbon monoxide can cause fainting, coma or even death.

Sulfur dioxide is produced when fuels that contain sulfur impurities are burned.

Sulfur dioxide can mix with water in clouds to produce sulfuric acid, so cause acid rain.

Sulfur dioxide can cause respiratory problems.

Nitrogen oxides are produced when nitrogen and oxygen from the air react/combine due to the heat of burning.

Nitrogen oxides can mix with water in clouds to produce nitric acid, so cause acid rain.

Nitrogen oxides can cause respiratory problems.

12.1 Iron, tin, copper *[2 marks if all correct, or 1 mark if 1 correct]*

12.2 $2AgCl_{(aq)} + Cu_{(s)} \rightarrow CuCl_{2(aq)} + 2Ag_{(s)}$ *[1 mark for correct equation, 1 mark for balancing, 1 mark for state symbols]*

13.1 Copper is lower in the reactivity series/less reactive than carbon *[1 mark]*, so can be extracted by reduction using carbon *[1 mark]*.

13.2 Bacteria are used to convert copper compounds in the ore into soluble copper compounds *[1 mark]*. This produces a leachate that contains copper ions *[1 mark]* which can be extracted by electrolysis/displacement with iron *[1 mark]*.

13.3 The atoms in copper form layers which slide over each other, so it can be drawn out into wires *[1 mark]*. Copper contains delocalised electrons which are free to move and carry an electric current *[1 mark]*.

13.4 The tin atoms in bronze distort the structure of the copper *[1 mark]*. This means the layers can no longer slide over each other *[1 mark]*, so bronze is harder than copper *[1 mark]*.

14.1 Polymer **A** has weak forces between the chains *[1 mark]*. Polymer **B** has cross links between the chains *[1 mark]*.

14.2 Polymer **B** *[1 mark]* as it's rigid, so would keep the shape of the mug *[1 mark]* and it wouldn't be softened by the hot drinks *[1 mark]*.

15.1
```
     H  H
     |  |
 H—C—C—O—H
     |  |
     H  H
```
[1 mark for correct number of each atom, 1 mark for atoms joined up correctly]

15.2 Ethanol can be made by reacting ethene with steam *[1 mark]* in the presence of a catalyst *[1 mark]*. It can also be made by fermenting sugars with yeast *[1 mark]* at around 37 °C *[1 mark]* and slightly acidic conditions *[1 mark]* in the absence of oxygen *[1 mark]*.

16.1 Add a few drops of sodium hydroxide to a sample of the solution *[1 mark]*. If iron(II) ions are present, a green precipitate should form *[1 mark]*.
$FeSO_{4(aq)} + 2NaOH_{(aq)} \rightarrow Fe(OH)_{2(s)} + Na_2SO_{4(aq)}$ *[1 mark for correct equation, 1 mark for balancing, 1 mark for state symbols]*

16.2 Add some barium chloride *[1 mark]* to the solution in the presence of hydrochloric acid *[1 mark]*. If sulfate ions are present, a white precipitate should form *[1 mark]*.
$BaCl_{2(aq)} + FeSO_{4(aq)} \rightarrow BaSO_{4(s)} + FeCl_{2(aq)}$ *[1 mark for balanced equation, 1 mark for state symbols]*

16.3 Add iron(II) oxide to sulfuric acid until the reaction stops / the solid sinks to the bottom *[1 mark]*. Filter off the excess iron(II) oxide *[1 mark]*. Gently heat the iron(II) sulfate solution to evaporate some of the water and then leave to cool *[1 mark]*. Filter and dry the crystals that form *[1 mark]*.

17.1 Number of moles of carbon = mass ÷ A_r = 24 ÷ 12 = 2 mol
1 mol of carbon reacts to produce 2 mol of hydrogen gas, so 2 mol of carbon will react to produce 2 × 2 = 4 mol of hydrogen gas.
M_r of H_2 = 2 × 1 = 2
Mass = number of moles × molar mass = 4 × 2 = **8 g**
[4 marks for correct answer, or 1 mark for correct number of moles of carbon, 1 mark for correct number of moles of hydrogen, 1 mark for correct M_r of hydrogen]

17.2 1 mole of gas occupies 24 dm³ at room temperature and pressure, so 4 moles of gas occupies 4 × 24 = **96 dm³** *[1 mark]*
If you got the answer to 13.1 wrong you still get the mark here if you used your answer to 13.1 correctly in this part.

17.3 Percentage yield
$= \dfrac{\text{Mass of product actually made}}{\text{Maximum theoretical mass of product}} \times 100 = \dfrac{4.8}{8} \times 100$
= **60%** *[2 marks for correct answer, otherwise 1 mark for using the correct equation to calculate percentage yield]*

18.1 100% *[1 mark]*. The reaction only has one product *[1 mark]*.

18.2 E.g. the reaction is reversible / some of the products will always turn back into reactants / there might be side reactions / some of the product may be lost as it's separated from the reaction mixture *[1 mark]*.

18.3 A low temperature shifts the position of equilibrium in favour of the forward, exothermic reaction *[1 mark]*. This means there will be more product at equilibrium / the yield will be greater *[1 mark]*. However, a low temperature decreases the rate of reaction *[1 mark]*. So the temperature is a compromise in order to get a good yield at a reasonable rate *[1 mark]*.

19.1 $H^+ + OH^- \rightarrow H_2O$ *[1 mark]*

19.2 sodium sulfate *[1 mark]*

19.3 Universal indicator doesn't have a sudden colour change at the endpoint *[1 mark]*. An indicator such as methyl orange / phenolphthalein / litmus *[1 mark]* should be used instead.

19.4 volume in dm³ = 20.35 ÷ 1000 = 0.02035 dm³
moles = volume × concentration = 0.02035 × 0.10 = **0.0020 mol** *[3 marks for correct answer, otherwise 1 mark for volume in dm³, 1 mark for using the correct equation to calculate moles]*

19.5 mean titre = (20.05 + 19.95 + 20.00) ÷ 3 = **20.00 cm³**
[2 marks for correct answer, otherwise 1 mark for ignoring rough titre]

19.6 volume of H_2SO_4 to react in dm³ = 20.0 ÷ 1000 = 0.0200 dm³
moles of H_2SO_4 to react = 0.0200 × 0.10 = 0.0020 mol
$H_2SO_4 + 2NaOH \rightarrow Na_2SO_4 + 2H_2O$
1 mole of H_2SO_4 reacts with 2 moles of NaOH, so 0.0020 mol of H_2SO_4 reacts with 0.0040 mol of NaOH
Concentration of NaOH = number of moles ÷ volume
= 0.0040 ÷ 0.025 = **0.16 mol/dm³** *[5 marks for correct answer, otherwise 1 mark for number of moles of H_2SO_4, 1 mark for number of moles of NaOH, 1 mark for balanced reaction equation, 1 mark for using the correct equation to calculate concentration]*

20.1   Aluminium ore is mixed with cryolite and melted *[1 mark]*. An electric current is passed through the molten ore *[1 mark]*. At the cathode/negative electrode, Al^{3+} ions are reduced to aluminium metal *[1 mark]*: $Al^{3+} + 3e^- \rightarrow Al$ *[1 mark]*. At the anode/positive electrode, O^{2-} ions are oxidised to oxygen *[1 mark]*: $2O^{2-} \rightarrow O_2 + 4e^-$ *[1 mark]*.

20.2   When aluminium corrodes it forms a protective layer of aluminium oxide *[1 mark]* that stops any further reaction taking place *[1 mark]*.

20.3   Galvanising means covering iron with a layer of zinc *[1 mark]*. This acts as a protective barrier to keep out water and oxygen *[1 mark]*. If the layer is scratched, the zinc around the scratch reacts instead of the iron *[1 mark]*.

21.1   Order: diamond, poly(propene), butane *[1 mark]*. Explanation: Diamond has the highest melting point as you need to break the strong covalent bonds *[1 mark]*. Poly(propene) forms larger molecules than butane, so has stronger intermolecular forces *[1 mark]*, which require more energy to break *[1 mark]*.

21.2   Particles with a diameter between 1 nm and 100 nm / particles containing only a few hundred atoms *[1 mark]*. They have a very high surface area to volume ratio compared to bulk materials *[1 mark]*.

21.3   E.g. they could be used for drug delivery *[1 mark]*. The effects of nanoparticles on health aren't understood / they could react with things in the body / they could damage cells *[1 mark]*.

22.1   Zinc is more reactive than hydrogen *[1 mark]*. This means zinc forms positive ions more easily than hydrogen *[1 mark]*.

22.2   Reduction *[1 mark]*, because the hydrogen ions gain electrons *[1 mark]*.

22.3 $4OH^- \rightarrow O_2 + 2H_2O + 4e^-$ *[1 mark for correct reactants and products, 1 mark for balancing]*

If you had '–4e⁻' on the left hand side of the equation instead of '+4e⁻' on the right, you still get the marks.

23.1   Similarity: e.g. they both form positive ions / they both react with acid *[1 mark]*. Difference: e.g. cobalt has a higher melting point / cobalt forms more than one positive ion / cobalt reacts less vigorously with acid *[1 mark]*.

23.2   How to grade your answer:

Level 0:   There is no relevant information. *[No marks]*

Level 1:   There is a brief description of the similarities and differences between lithium and sodium, but no explanation of these observations.
           [1 to 2 marks]

Level 2:   There is a detailed comparison of the similarities and differences between lithium and sodium, and some explanation of the observations.
           [3 to 4 marks]

Level 3:   There is a detailed comparison of the similarities and differences between lithium and sodium, and a good explanation of the observations.
           [5 to 6 marks]

Here are some points your answer may include:

Both react to form positive, 1+ ions.

Both elements are in Group 1, so have one electron in their outer shell.

Not much energy is needed to remove this one outer electron and give the elements a full outer shell of electrons.

Both react with acid.

Sodium reacts more vigorously with acid than lithium.

Sodium is lower down in the group, so the outer electron in sodium is further away from the nucleus than the outer electron in lithium.

The attraction between the outer electron and the nucleus of sodium is less than the attraction between the outer electron and the nucleus in lithium.

Less energy is needed to remove the outer electron of sodium, making it more reactive than lithium.

23.3   Any answer in the range 80–160 °C *[1 mark]*.

Answers